Overview

"The Hidden Horizon: Revealing the Reality of our Round Earth" is a comprehensive exploration of the flat earth theory and its debunking. In this book, we delve into the historical background of the flat earth theory, examine the key beliefs of flat earthers, and provide scientific evidence against this theory. We also investigate the mysterious ice wall, exploring legends and myths surrounding it, scientific research and discoveries in the Antarctic region, and debunking claims about its existence. Through a thorough examination of gravity, the Coriolis effect, the curvature of the earth, and the round earth in space, we present a compelling case for the reality of our round earth. Furthermore, this book delves into the role of science and the scientific method in understanding the shape of our planet. We explore cognitive biases, belief systems, and psychological factors that contribute to the flat earth movement, while debunking these psychological factors. Additionally, we address the challenges in communicating complex science and provide effective science communication strategies to promote scientific literacy. We also discuss the societal and educational consequences of flat earth beliefs, the dangers of ignoring scientific consensus, and the importance of promoting critical thinking and rationality. In conclusion, "The Hidden Horizon" emphasizes the importance of embracing the reality of our round earth and moving forward in the age of information..

Table Of Contents

1 Introduction to the Flat Earth Theory

1.1 Historical Background of the Flat Earth Theory

The belief in a flat Earth has a long and complex history, dating back to ancient civilizations. While it may seem surprising to us today, the idea of a flat Earth was widely accepted for centuries before scientific evidence proved otherwise. In this section, we will explore the historical background of the flat Earth theory, examining its origins and the factors that contributed to its widespread acceptance.

Ancient Civilizations and Early Beliefs

The concept of a flat Earth can be traced back to ancient civilizations such as Mesopotamia, Egypt, and Greece. In these early societies, people observed the world around them and developed their own explanations for its shape. Due to limited technology and knowledge, their understanding of the Earth's shape was based on their immediate surroundings and personal experiences.

In Mesopotamia, for example, the belief in a flat Earth was influenced by the geography of the region. The flat and fertile plains of the Tigris and Euphrates rivers led people to perceive the Earth as a flat disc floating on water. Similarly, in ancient Egypt, the Nile River's annual flooding created a flat and fertile floodplain, reinforcing the idea of a flat Earth.

Mythology and Religious Beliefs

Religion and mythology also played a significant role in shaping the belief in a flat Earth. Many ancient cultures had creation myths that depicted the Earth as a flat, disc-like structure. These myths often involved gods or divine beings who created the Earth and placed it on a flat surface.

In Norse mythology, for instance, the Earth was believed to be a flat disc surrounded by a vast ocean. The gods created a protective barrier in the form of a giant serpent, known as Jormungandr, which encircled the Earth, preventing it from falling into the abyss. Similar beliefs can be found in other mythologies, where the Earth was often depicted as a flat plane with a dome-like sky above.

Influence of Philosophers and Scholars

During the classical period, Greek philosophers and scholars made significant contributions to the understanding of the Earth's shape. However, even some renowned thinkers of the time, such as Pythagoras and Parmenides, believed in a flat Earth.

Pythagoras, the famous mathematician and philosopher, proposed a spherical Earth, but his followers rejected this idea and continued to advocate for a flat Earth. Parmenides, on the other hand, argued that the Earth was a flat disk floating in an infinite sea.

The influential Greek philosopher Plato also contributed to the belief in a flat Earth. In his work "Timaeus," Plato described the Earth as a flat, circular plane divided into concentric circles. This influential text further solidified the notion of a flat Earth in the minds of many.

Medieval Period and the Church's Influence

During the medieval period, the belief in a flat Earth became deeply ingrained in European society, largely due to the influence of the Christian Church. The Church, as the dominant institution of the time, played a significant role in shaping people's beliefs and controlling knowledge.

The Church's interpretation of biblical texts, such as the Book of Isaiah and the Book of Revelation, was often used to support the idea of a flat Earth. These texts were taken literally, and any suggestion of a spherical Earth was considered heretical.

Additionally, the Church's authority and control over education limited the dissemination of alternative ideas. Scholars who challenged the prevailing belief in a flat Earth risked being labeled as heretics and facing severe consequences.

Renaissance and the Emergence of Scientific Inquiry

The Renaissance period marked a significant shift in the understanding of the Earth's shape. As scientific inquiry and exploration expanded, new evidence began to challenge the prevailing belief in a flat Earth.

One of the key figures in this shift was the Italian astronomer and mathematician Galileo Galilei. Through his observations of celestial bodies and his development of the telescope, Galileo provided compelling evidence for a spherical Earth. His discoveries, along with those of other scientists and explorers, gradually eroded the belief in a flat Earth.

Conclusion

The historical background of the flat Earth theory reveals the complex interplay between cultural beliefs, religious interpretations, and limited scientific knowledge. Over time, as scientific advancements and empirical evidence accumulated, the belief in a flat Earth was gradually debunked.

In the next section, we will delve into the key beliefs of flat Earthers, examining the modern manifestation of this theory and the arguments put forth by its proponents.

1.2 Key Beliefs of Flat Earthers

Flat Earth theory is a controversial belief system that asserts that the Earth is flat rather than spherical. While the overwhelming scientific consensus supports the idea that the Earth is round, Flat Earthers hold a number of key beliefs that challenge this widely accepted view. In this section, we will explore some of the main beliefs held by Flat Earthers and examine the evidence that contradicts these claims.

1.2.1 Earth's Flat Surface

One of the fundamental beliefs of Flat Earthers is that the Earth is a flat, disc-shaped plane. They argue that the curvature of the Earth is not observable in everyday life and that any perceived curvature is simply an optical illusion. Flat Earthers often claim that the horizon appears flat and level, suggesting that the Earth is not curved.

However, scientific evidence overwhelmingly supports the idea that the Earth is a sphere. The curvature of the Earth can be observed in various ways, such as the way ships disappear over the horizon or the way the tops of distant buildings become visible as you approach them. Additionally, the phenomenon of Earth's shadow during a lunar eclipse provides further evidence of a spherical Earth.

1.2.2 The Firmament and the Dome

Another belief held by some Flat Earthers is the existence of a firmament or dome above the Earth. They argue that this dome acts as a barrier, separating the Earth from the rest of the universe. According to this belief, the stars, moon, and sun are all contained within this dome and move in a circular motion above the flat Earth.

However, modern astronomy and space exploration have provided ample evidence to refute this claim. Astronomical observations and space missions have shown that the Earth is just one of many celestial bodies in the vast universe. The movement of celestial objects can be explained by the laws of physics and the principles of orbital mechanics, rather than the existence of a firmament.

1.2.3 Gravity and the Flat Earth

Flat Earthers often question the concept of gravity as it is understood by the scientific community. They propose alternative explanations for the force that holds objects to the Earth's surface, such as the idea of density and buoyancy. According to this belief, objects are pushed down towards the Earth's surface due to their relative density compared to the surrounding air or water.

However, the concept of gravity has been extensively studied and verified through numerous experiments and observations. Gravity is a fundamental force that explains the motion of celestial bodies, the shape of the Earth, and the behavior of objects on its surface. The theory of gravity is supported by a vast body of scientific evidence and is widely accepted by the scientific community.

1.2.4 Conspiracy Theories and Skepticism

Flat Earthers often express skepticism towards mainstream science and institutions, believing that there is a global conspiracy to hide the true nature of the Earth. They argue that governments, space agencies, and scientists are all part of a grand cover-up to maintain the illusion of a round Earth.

However, the idea of a global conspiracy involving millions of people across different countries and organizations is highly implausible. It would require an extraordinary level of coordination and secrecy, which is simply not feasible. Furthermore, the scientific method encourages skepticism and critical thinking, and scientific knowledge is built upon rigorous testing and peer review.

In conclusion, Flat Earthers hold a number of key beliefs that challenge the widely accepted view of a round Earth. However, these beliefs are not supported by scientific evidence and are contradicted by a vast body of research and observations. The overwhelming consensus among scientists and experts is that the Earth is indeed a sphere. In the following sections, we will delve deeper into the scientific evidence that supports the round Earth theory and debunk the claims made by Flat Earthers.

1.3 Common Arguments and Debunking

In this section, we will explore some of the common arguments put forth by flat earthers and provide scientific evidence to debunk these claims. It is important to address these arguments and provide a clear understanding of the reality of our round Earth.

1.3.1 The Horizon Always Appears Flat

One of the main arguments made by flat earthers is that the horizon always appears flat, therefore suggesting a flat Earth. However, this argument fails to take into account the limitations of human perception and the vastness of the Earth.

When we observe the horizon from ground level, it may indeed appear flat to the naked eye. However, as we gain altitude, such as from an airplane or a high vantage point, the curvature of the Earth becomes more apparent. This is because the Earth's curvature is gradual and not easily noticeable from ground level.

Furthermore, the curvature of the Earth can be observed in various ways. For example, when ships sail away from the shore, they gradually disappear from view, with the hull disappearing first and then the mast. This phenomenon, known as the "ship disappearing over the horizon," is a clear indication of the Earth's curvature.

1.3.2 Lack of Visible Curvature in Everyday Life

Another argument put forth by flat earthers is the perceived lack of visible curvature in everyday life. They claim that if the Earth were round, we would be able to see the curvature from high buildings or mountains.

However, the curvature of the Earth is not easily noticeable in our day-to-day lives due to its large size. The Earth has a circumference of approximately 40,075 kilometers (24,901 miles), which means that the curvature is relatively subtle. To put it into perspective, the curvature of the Earth would only be about 8 inches per mile.

When standing on the ground, the curvature is not readily apparent because our field of view is limited. It is only when we have a wide panoramic view, such as from an airplane or a high-altitude photograph, that the curvature becomes more evident.

1.3.3 Gravity and the Flat Earth

Flat earthers often question the concept of gravity and argue that it is not a force but rather a result of density and buoyancy. They claim that objects fall to the ground because they are denser than the air around them, rather than being pulled towards the center of the Earth.

However, the scientific understanding of gravity is well-established and supported by extensive evidence. Gravity is a fundamental force that attracts objects with mass towards each other. It is responsible for holding celestial bodies, such as planets and moons, in their orbits.

Experiments and observations have consistently demonstrated the effects of gravity. For example, when objects are dropped from a height, they fall towards the ground in a predictable manner, regardless of their density. Additionally, the phenomenon of weightlessness experienced by astronauts in space is a direct result of the absence of gravity.

1.3.4 The Sun and Moon's Movement

Flat earthers often question the movement of the Sun and Moon in the sky, suggesting that their paths are inconsistent with a round Earth. They argue that if the Earth were round, the Sun and Moon would always appear directly above the observer.

In reality, the movement of the Sun and Moon can be explained by the Earth's rotation and its tilted axis. As the Earth rotates on its axis, different parts of the planet are exposed to sunlight, resulting in day and night. The apparent path of the Sun across the sky, known as the ecliptic, is a result of the Earth's tilt and its orbit around the Sun.

The Moon's movement is also consistent with a round Earth. Its phases, as observed from Earth, are a result of the relative positions of the Sun, Earth, and Moon. The Moon's orbit around the Earth causes it to appear to change shape throughout the lunar month.

1.3.5 Lack of Photos of the Earth from Space

Flat earthers often claim that the lack of photographs of the entire Earth from space is evidence of a conspiracy to hide the true shape of the Earth. They argue that if the Earth were round, there should be numerous images of the entire planet.

In reality, there are countless photographs of the Earth from space, taken by satellites, astronauts, and space probes. These images clearly show the Earth as a round, spherical object. The most famous of these images is the "Blue Marble" photograph taken by the Apollo 17 mission in 1972, which shows the Earth in its entirety.

Furthermore, the International Space Station (ISS) orbits the Earth and provides a continuous stream of live video footage, showing the Earth's curvature and the transition from day to night. The existence of these images and videos debunks the notion that there is a conspiracy to hide the true shape of the Earth.

In conclusion, the common arguments put forth by flat earthers can be easily debunked with scientific evidence. The curvature of the Earth, the effects of gravity, the movement of the Sun and Moon, and the abundance of photographs from space all provide irrefutable proof of our round Earth. It is important to critically evaluate these arguments and embrace the reality of our planet's shape.

1.4 Scientific Evidence Against the Flat Earth Theory

The flat earth theory has gained some traction in recent years, with proponents arguing that the earth is a flat disc rather than a spherical shape. However, when we examine the scientific evidence, it becomes clear that the flat earth theory is not supported by empirical data or rigorous scientific research. In this section, we will explore some of the key scientific evidence that debunks the flat earth theory and supports the reality of our round earth.

1.4.1 The Shape of Earth's Shadow during Lunar Eclipses

One of the most compelling pieces of evidence against the flat earth theory comes from the observation of lunar eclipses. During a lunar eclipse, the earth casts a shadow on the moon, creating a reddish hue. If the earth were flat, the shadow would appear as a straight line across the moon's surface. However, what we observe is a curved shadow that matches the shape of a sphere. This phenomenon can only be explained if the earth is indeed a sphere.

1.4.2 The Curvature of the Horizon

Another piece of evidence that supports the round earth theory is the curvature of the horizon. As we observe the horizon from different vantage points, such as from the top of a tall building or from an airplane, we notice that the horizon appears curved. This curvature is consistent with the predictions of a spherical earth and cannot be explained by a flat earth model. Additionally, the degree of curvature observed aligns with the mathematical calculations based on the earth's radius.

1.4.3 Gravity and the Shape of Water

Gravity plays a crucial role in shaping the earth and its bodies of water. On a flat earth, bodies of water would naturally seek a level surface, resulting in a perfectly flat ocean. However, we observe that water bodies, such as oceans and seas, have a curved shape. This curvature is a direct consequence of gravity, which pulls the water towards the center of the earth, forming a spherical shape. The fact that bodies of water exhibit this curvature provides strong evidence for the round earth theory.

1.4.4 Circumnavigation and Flight Paths

The ability to circumnavigate the globe is another piece of evidence that supports the round earth theory. When we examine flight paths and navigation routes, we find that they are based on the assumption of a spherical earth. For example, flights from New York to Tokyo often take a polar route, flying over the Arctic region. This route makes sense on a round earth, as it is the shortest distance between the two points. On a flat earth, however, this flight path would be illogical and inefficient. The fact that we can successfully navigate the globe using these flight paths is strong evidence against the flat earth theory.

1.4.5 Satellite Imagery and Space Exploration

Satellite imagery and space exploration have provided us with a wealth of evidence supporting the round earth theory. Satellites orbiting the earth capture images of our planet from space, clearly showing its spherical shape. These images are not only visually striking but also consistent with the mathematical models and calculations based on a round earth. Furthermore, space missions, such as those conducted by NASA and other space agencies, have provided us with direct observations and measurements that confirm the earth's spherical shape.

1.4.6 Gravity's Effect on Time

Another fascinating piece of evidence against the flat earth theory comes from the phenomenon of time dilation. According to Einstein's theory of general relativity, gravity affects the passage of time. This effect has been observed and measured through experiments and satellite technology. For example, GPS satellites, which rely on precise timing, must account for the difference in time due to the gravitational pull of the earth. If the earth were flat, we would not observe these time discrepancies, providing further evidence for the round earth theory.

In conclusion, the scientific evidence overwhelmingly supports the reality of our round earth. From the shape of the earth's shadow during lunar eclipses to the curvature of the horizon and the observations from satellite imagery and space exploration, all the evidence points to a spherical earth. The flat earth theory lacks empirical support and fails to explain the multitude of observations and measurements that confirm the round earth model. As we delve deeper into the scientific evidence, it becomes clear that the flat earth theory is nothing more than a misconception that can be debunked through rigorous scientific scrutiny.

2 The Shape of Our Planet

2.1 Early Observations and Theories

Throughout history, humans have been fascinated by the shape of the Earth and have made various observations and theories to understand its true form. Early civilizations, such as the ancient Greeks and Egyptians, were among the first to propose theories about the shape of our planet. These early observations and theories laid the foundation for our modern understanding of Earth's shape.

2.1.1 Ancient Greek Contributions

One of the earliest recorded theories about the shape of the Earth comes from ancient Greece. Around the 6th century BCE, Greek philosophers such as Pythagoras and Parmenides proposed that the Earth was a sphere. They based their theories on several observations, including the circular shadow cast by the Earth during a lunar eclipse and the way ships disappeared over the horizon.

The most influential Greek philosopher, Aristotle, further supported the idea of a spherical Earth in the 4th century BCE. He observed that as one traveled further from their home, they would see different constellations in the night sky. This observation led him to conclude that the Earth must be curved.

2.1.2 Eratosthenes and the Measurement of Earth's Circumference

In the 3rd century BCE, the Greek mathematician and astronomer Eratosthenes made a significant contribution to our understanding of Earth's shape. By observing the angle of the Sun's rays at different locations, he was able to estimate the Earth's circumference with remarkable accuracy.

Eratosthenes noticed that at noon on the summer solstice, the Sun was directly overhead in the Egyptian city of Syene (modern-day Aswan), casting no shadow. However, in Alexandria, located north of Syene, he observed that a vertical object cast a shadow. By measuring the angle of the shadow and knowing the distance between the two cities, Eratosthenes calculated the Earth's circumference to be approximately 39,375 kilometers (24,662 miles). This estimation was remarkably close to the actual value.

2.1.3 Islamic Contributions

During the Islamic Golden Age (8th to 14th centuries CE), Muslim scholars made significant contributions to the understanding of Earth's shape. Islamic astronomers, such as Al-Biruni and Al-Farghani, built upon the works of the ancient Greeks and conducted their own observations and calculations.

Al-Biruni, an 11th-century Persian scholar, proposed a method to calculate the Earth's radius using trigonometry. He measured the angle between the horizon and the peak of a mountain and used this information to estimate the Earth's radius accurately.

2.1.4 Ferdinand Magellan's Circumnavigation

In the 16th century, the Portuguese explorer Ferdinand Magellan embarked on a voyage that would provide further evidence for the spherical shape of the Earth. Magellan's expedition was the first to circumnavigate the globe, proving that the Earth was not flat.

As Magellan and his crew sailed westward, they observed that the stars shifted in the night sky, confirming that they were moving along a curved path. Additionally, they noticed that the horizon appeared to curve as they traveled, further supporting the idea of a spherical Earth.

2.1.5 Gravity and Earth's Shape

In the 17th century, the English scientist Sir Isaac Newton formulated the theory of gravity, which provided a deeper understanding of Earth's shape.

Newton's law of universal gravitation explained how objects are attracted to one another based on their mass and distance.

The force of gravity acts towards the center of mass of an object, causing it to form a spherical shape. This principle applies to celestial bodies, including Earth. The gravitational force pulls matter towards the center, resulting in a spherical shape.

2.1.6 Modern Geodetic Surveys and Satellite Measurements

In recent centuries, advancements in technology have allowed for more accurate measurements of Earth's shape. Geodetic surveys, which involve precise measurements of the Earth's surface, have provided valuable data on the planet's curvature.

Satellite measurements have also played a crucial role in confirming the spherical shape of the Earth. Satellites orbiting the planet capture images of Earth from space, clearly showing its round shape. These images provide irrefutable evidence that the Earth is not flat.

In conclusion, early observations and theories, along with modern scientific advancements, have overwhelmingly disproven the flat Earth theory. From the ancient Greeks to modern satellite imagery, the evidence for a spherical Earth is abundant and conclusive. The understanding of Earth's shape has evolved over time, thanks to the contributions of numerous scientists and explorers.

2.2 Modern Understanding of Earth's Shape

The shape of our planet has been a subject of fascination and exploration for centuries. Early observations and theories laid the foundation for our modern understanding of Earth's shape. Through geodetic surveys, satellite imagery, and space exploration, we have gathered overwhelming evidence that supports the fact that our planet is round. In this section, we will delve into the modern understanding of Earth's shape and the scientific evidence that debunks the flat earth theory.

The Spherical Earth

The concept of a spherical Earth dates back to ancient civilizations such as the Greeks and Egyptians. Early astronomers and mathematicians, including Pythagoras and Eratosthenes, made significant contributions to our understanding of Earth's shape. They observed the curvature of the Earth's shadow during lunar eclipses and measured the angle of the Sun's rays at different locations on Earth. These observations led them to conclude that the Earth is a sphere.

Gravity and Earth's Shape

One of the key factors that contribute to Earth's shape is gravity. Gravity is the force that attracts objects towards the center of mass. On a spherical Earth, gravity pulls everything towards the center, creating a balanced distribution of mass. This gravitational force acts equally in all directions, resulting in a spherical shape.

Geodetic Surveys and Measurements

Geodetic surveys play a crucial role in determining the shape of our planet. These surveys involve precise measurements of Earth's surface and its gravitational field. By using advanced technologies such as GPS (Global

Positioning System) and satellite imagery, scientists can accurately measure the Earth's curvature and shape.

Geodetic measurements have revealed that Earth is not a perfect sphere but an oblate spheroid. This means that the Earth is slightly flattened at the poles and bulges at the equator. The equatorial diameter is larger than the polar diameter, resulting in a shape similar to that of a squashed ball.

Satellite Imagery and Space Exploration

The advent of satellite imagery and space exploration has provided us with undeniable evidence of Earth's roundness. Satellites orbiting the Earth capture images that clearly show the curvature of our planet. These images, taken from various angles and altitudes, offer a comprehensive view of Earth's shape.

Space missions, such as those conducted by NASA and other space agencies, have further confirmed the roundness of Earth. Astronauts aboard the International Space Station (ISS) have witnessed the curvature of Earth firsthand. They have shared breathtaking photographs and videos that showcase the beauty and roundness of our planet.

Debunking the Flat Earth Theory

Despite the overwhelming evidence supporting the roundness of Earth, the flat earth theory continues to persist among a small group of individuals. However, their arguments and claims can be easily debunked using scientific reasoning and evidence.

One of the common arguments put forth by flat earthers is the belief that the horizon always appears flat. However, this is a result of the limited perspective of the human eye. As we increase our altitude or use telescopes and cameras with zoom capabilities, we can observe the curvature of the Earth.

Another claim made by flat earthers is the existence of an ice wall surrounding the Earth. They argue that this ice wall prevents us from falling off the edge of the flat Earth. However, there is no scientific evidence to support the existence of such an ice wall. Explorations of the Antarctic region have revealed vast expanses of ice, but there is no continuous wall encircling the Earth.

In conclusion, the modern understanding of Earth's shape is based on centuries of scientific observations, geodetic surveys, satellite imagery, and space exploration. The evidence overwhelmingly supports the fact that our planet is round, debunking the flat earth theory. The spherical shape of Earth is a fundamental aspect of our understanding of the natural world, and it is crucial to embrace this reality to foster scientific literacy and critical thinking.

2.3 Geodetic Surveys and Measurements

Geodetic surveys and measurements play a crucial role in understanding the shape of our planet. These scientific techniques have been used for centuries to gather data and provide evidence that supports the round Earth theory. By examining the Earth's curvature and measuring its dimensions, geodetic surveys have consistently debunked the claims of the flat Earth theory.

2.3.1 Early Geodetic Surveys

Early geodetic surveys were conducted by ancient civilizations to determine the size and shape of the Earth. One of the most notable examples is the work of the ancient Greek philosopher and mathematician, Eratosthenes. In the 3rd century BCE, Eratosthenes calculated the Earth's circumference using simple trigonometry and measurements of the angle of the Sun's rays at two different locations. His calculations were remarkably accurate, providing strong evidence for a spherical Earth.

Throughout history, geodetic surveys have been refined and improved, utilizing advanced instruments and techniques. These surveys involve measuring the Earth's surface and its gravitational field to determine its shape and dimensions.

2.3.2 Modern Geodetic Surveys

In modern times, geodetic surveys have become even more precise and sophisticated. The development of advanced technologies, such as Global Navigation Satellite Systems (GNSS), has revolutionized the field of geodesy. GNSS systems, like the Global Positioning System (GPS), allow for highly accurate measurements of positions on the Earth's surface.

Geodetic surveys involve measuring the Earth's curvature, which is a fundamental characteristic of a round Earth. By using precise instruments and mathematical calculations, scientists can determine the curvature of the Earth over large distances. These measurements consistently show that the Earth is curved, providing strong evidence against the flat Earth theory.

2.3.3 Measuring Earth's Curvature

One of the key methods used in geodetic surveys is measuring the curvature of the Earth. This can be done through various techniques, including leveling, triangulation, and satellite-based measurements.

Leveling involves measuring the height differences between different points on the Earth's surface. By comparing these height differences over a known distance, scientists can calculate the curvature of the Earth. These measurements consistently show that the Earth's surface curves away from a flat plane.

Triangulation is another method used to measure the Earth's curvature. It involves measuring the angles between three or more points on the Earth's surface and using trigonometry to calculate distances and curvature. This technique has been used extensively in geodetic surveys and has provided further evidence for a round Earth.

Satellite-based measurements, such as those obtained from GNSS systems, also contribute to our understanding of the Earth's curvature. Satellites orbiting the Earth provide precise measurements of positions, which can be used to calculate the curvature of the Earth. These measurements align with the findings from other geodetic surveys, reinforcing the evidence for a round Earth.

2.3.4 Determining Earth's Dimensions

Geodetic surveys not only measure the curvature of the Earth but also provide valuable data on its dimensions. By measuring distances between points on the

Earth's surface and using mathematical calculations, scientists can determine the Earth's radius, circumference, and other important measurements.

One of the most famous geodetic surveys was conducted by the French Academy of Sciences in the 18th century. This survey, known as the Meridian Arc Measurement, aimed to accurately measure the length of a meridian arc from Dunkirk in northern France to Barcelona in Spain. The results of this survey provided valuable data for calculating the Earth's dimensions and further supported the round Earth theory.

Modern geodetic surveys continue to refine our understanding of the Earth's dimensions. By combining data from various sources, including satellite measurements and ground-based surveys, scientists can calculate the Earth's radius, circumference, and shape with remarkable accuracy.

2.3.5 Consistency of Geodetic Surveys

One of the most compelling aspects of geodetic surveys is their consistency. Measurements and calculations conducted by different scientists and organizations around the world consistently yield results that support the round Earth theory. This consistency across different methods and locations provides strong evidence against the flat Earth theory.

Geodetic surveys and measurements have been conducted by scientists from various countries and backgrounds, all reaching the same conclusion: the Earth is round. These surveys have been peer-reviewed, scrutinized, and replicated by independent researchers, further reinforcing the validity of their findings.

In conclusion, geodetic surveys and measurements have played a crucial role in disproving the flat Earth theory. Through the use of advanced instruments, mathematical calculations, and satellite-based measurements, scientists have consistently demonstrated the curvature of the Earth and determined its dimensions. The evidence gathered from geodetic surveys supports the overwhelming consensus among scientists that our planet is indeed round.

2.4 Satellite Imagery and Space Exploration

Satellite imagery and space exploration have played a crucial role in disproving the flat earth theory and revealing the true shape of our planet. Over the years, advancements in technology have allowed us to capture stunning images of Earth from space, providing undeniable evidence of its spherical nature. In this section, we will explore how satellite imagery and space exploration have contributed to our understanding of Earth's shape and debunked the claims of flat earthers.

2.4.1 The Power of Satellite Imagery

Satellite imagery has revolutionized our perception of Earth. Satellites orbiting our planet capture high-resolution images that showcase the curvature of Earth's surface. These images provide a comprehensive view of our planet, showing its spherical shape and the vastness of its oceans, continents, and atmosphere. The images obtained from satellites have become an essential tool for meteorologists, geographers, and scientists in various fields.

One of the most iconic satellite images is the "Blue Marble" photograph taken by the Apollo 17 mission in 1972. This image, showing Earth as a beautiful blue sphere floating in space, had a profound impact on our collective understanding of our planet's shape. Since then, numerous satellites, such as the Landsat series and the GOES satellites, have continuously captured images of Earth, further reinforcing the evidence of its roundness.

2.4.2 Earth Observation Satellites

In addition to capturing stunning images, Earth observation satellites provide valuable data that helps us understand our planet's dynamics. These satellites monitor various aspects of Earth, including weather patterns, climate change, and environmental conditions. By collecting data from different regions of the globe, these satellites contribute to our understanding of Earth's interconnected systems.

For example, weather satellites like the Geostationary Operational Environmental Satellites (GOES) provide real-time images and data on weather patterns, allowing meteorologists to track storms, monitor atmospheric conditions, and predict weather events. These satellites orbit at a specific altitude, enabling them to capture images of Earth's entire disk at regular intervals. The continuous stream of images from these satellites clearly shows the curvature of Earth and the rotation of weather systems around its axis.

2.4.3 Space Exploration and Astronaut Testimonies

Space exploration has also played a significant role in debunking the flat earth theory. Astronauts who have traveled to space have provided firsthand accounts and testimonies of Earth's spherical shape. They have witnessed the curvature of Earth with their own eyes and have shared their experiences with the world.

During the Apollo missions, astronauts had the opportunity to observe Earth from a unique vantage point. They saw the roundness of our planet and the thin layer of atmosphere that protects and sustains life. Their descriptions and photographs of Earth from space have been instrumental in dispelling the notion of a flat Earth.

Furthermore, space missions like the International Space Station (ISS) have provided continuous live footage of Earth from space. Astronauts aboard the ISS regularly share images and videos of Earth, showcasing its spherical shape and the breathtaking views of our planet from orbit. These images and videos serve as a powerful visual testament to the reality of Earth's roundness.

2.4.4 Debunking Flat Earth Claims

Despite the overwhelming evidence provided by satellite imagery and space exploration, some flat earthers continue to cling to their beliefs. They often dismiss satellite images as part of a global conspiracy or claim that they are doctored or manipulated. However, these claims are baseless and lack scientific credibility.

Satellite imagery is not limited to government agencies or space organizations. Many private companies and individuals have access to satellite data and can independently verify the images. Additionally, the sheer number of satellites orbiting Earth, operated by different countries and organizations, makes it highly unlikely that a global conspiracy could manipulate all the data and images.

Furthermore, the testimonies of astronauts who have traveled to space cannot be easily dismissed. These individuals undergo rigorous training and are highly knowledgeable about the technical aspects of space travel. Their accounts of Earth's roundness align with scientific understanding and are supported by the evidence provided by satellite imagery.

In conclusion, satellite imagery and space exploration have played a pivotal role in disproving the flat earth theory. The images captured by satellites clearly show the curvature of Earth, while the testimonies of astronauts provide firsthand accounts of its spherical shape. Despite the overwhelming evidence, some individuals continue to hold onto their flat earth beliefs. However, their claims lack scientific credibility and are easily debunked by the wealth of data and images obtained from satellites. The reality of our round Earth is firmly established through the advancements in satellite technology and space exploration.

3 The Ice Wall Mystery

3.1 Legends and Myths Surrounding the Ice Wall

The concept of an ice wall surrounding the Earth is a popular belief among flat earthers. According to this theory, the ice wall acts as a barrier, preventing people from falling off the edge of the flat Earth. This idea has gained traction due to its inclusion in various works of fiction and the spread of misinformation on the internet. However, when we examine the legends and myths surrounding the ice wall, we find that they are not supported by scientific evidence and are instead rooted in misconceptions and misinterpretations.

One of the most well-known legends associated with the ice wall is the belief that it is guarded by a secret organization or military force to prevent anyone from reaching it. This idea has been perpetuated in movies, books, and conspiracy theories, leading to a sense of mystery and intrigue. However, there is no credible evidence to support the existence of such a secret organization or military presence. It is important to distinguish between fiction and reality when evaluating these claims.

Another myth surrounding the ice wall is the idea that it stretches infinitely in all directions, creating an impenetrable barrier. This notion is often accompanied by claims that explorers who have attempted to reach the ice wall have mysteriously disappeared or been silenced. However, these stories are largely based on hearsay and lack verifiable evidence. In reality, the ice wall, known as the Antarctic ice sheet, is a vast expanse of ice that covers the continent of Antarctica. It is not an impenetrable barrier but rather a natural feature of our planet.

Furthermore, some flat earthers believe that the ice wall is the source of all freshwater on Earth. They argue that the melting ice from the wall replenishes the world's water supply. However, this claim is not supported by scientific understanding. The Earth's water cycle, which includes processes such as evaporation, condensation, and precipitation, is responsible for the distribution and replenishment of freshwater. The ice wall, while significant in size, does not play a central role in this natural process.

It is also worth noting that the concept of an ice wall surrounding a flat Earth is inconsistent with our understanding of gravity and the shape of our planet. Gravity, as explained by the theory of general relativity, causes matter to be attracted towards the center of mass. This gravitational force acts uniformly in all directions, resulting in a spherical shape for celestial bodies such as Earth. The presence of an ice wall surrounding a flat Earth would require a different explanation for gravity, which contradicts well-established scientific principles.

In contrast to the legends and myths surrounding the ice wall, scientific research and exploration have provided us with a wealth of knowledge about the Antarctic region. Numerous expeditions have been conducted to study the unique ecosystem, geology, and climate of Antarctica. These expeditions have revealed valuable information about the continent's history, including evidence of past climate change and the presence of ancient life forms.

Satellite imagery and remote sensing technologies have also allowed scientists to map and monitor the Antarctic ice sheet with great precision. These observations have provided valuable insights into the dynamics of ice flow, the effects of climate change on the ice sheet, and the contribution of Antarctica to global sea-level rise. The data collected from these studies have been instrumental in improving our understanding of Earth's climate system and its interconnected processes.

In conclusion, the legends and myths surrounding the ice wall are not supported by scientific evidence. They are often based on misconceptions, misinterpretations, and fictional narratives. Scientific research and exploration have provided us with a comprehensive understanding of the Antarctic region, revealing the true nature of the ice sheet and its significance in Earth's climate system. It is important to critically evaluate claims and rely on scientific evidence when examining such theories.

3.2 Exploring the Antarctic Region

The Antarctic region, with its vast expanse of ice and remote location, has long been a subject of fascination and mystery. It is also a key area of interest when it comes to debunking the flat earth theory. In this section, we will delve into the exploration of the Antarctic region and the scientific discoveries that have been made, shedding light on the reality of our round earth.

3.2.1 Early Expeditions and Discoveries

The exploration of the Antarctic region began in the early 19th century, with expeditions led by notable explorers such as James Cook, James Clark Ross, and Robert Falcon Scott. These early explorers ventured into the icy wilderness, braving harsh conditions and unknown dangers in their quest for knowledge.

One of the most significant discoveries made during these expeditions was the existence of the Antarctic ice sheet. These vast ice formations, covering the landmass of Antarctica, provided early evidence of the earth's curvature. As explorers traveled further south, they noticed that the horizon appeared to rise higher, indicating that they were standing on a curved surface.

3.2.2 Modern Scientific Research

In more recent times, scientific research in the Antarctic region has further solidified our understanding of the round earth. The International Geophysical Year (IGY) in 1957-1958 marked a significant milestone in Antarctic exploration, with numerous countries conducting research in various scientific disciplines.

One of the key findings from this period was the discovery of the Antarctic ozone hole. Scientists studying the ozone layer noticed a significant depletion of ozone over Antarctica, confirming the existence of a unique atmospheric phenomenon linked to the earth's curvature and the distribution of sunlight.

Furthermore, geodetic surveys and satellite measurements have provided precise data on the shape of the Antarctic continent. These measurements have revealed that the landmass is not a flat plane but rather a curved surface, consistent with the overall shape of the earth.

3.2.3 Ice Core Samples and Climate Research

Another crucial aspect of scientific research in the Antarctic region is the study of ice core samples. By drilling deep into the ice sheets, scientists can extract cylindrical samples that contain a historical record of the earth's climate.

These ice cores provide valuable insights into past climate patterns, atmospheric composition, and even the presence of ancient organisms. The analysis of these samples has revealed a wealth of information about the earth's history and its dynamic nature, further supporting the round earth model.

3.2.4 International Cooperation and Antarctic Treaties

The exploration and scientific research in the Antarctic region are not limited to a single country or organization. Instead, it is a collaborative effort involving multiple nations under the framework of international agreements.

The Antarctic Treaty System, established in 1959, sets out guidelines for the peaceful and scientific use of the region. It promotes international cooperation, sharing of scientific data, and the protection of the unique Antarctic environment. This global collaboration further reinforces the scientific consensus on the round earth, as researchers from different countries work together to uncover the truth.

3.2.5 Debunking Claims about the Ice Wall

One of the key claims made by flat earthers is the existence of an ice wall surrounding the earth's perimeter. According to this theory, the ice wall acts as a barrier, preventing people from reaching the edge of the flat earth.

However, the exploration of the Antarctic region has debunked this claim. Numerous expeditions and scientific research have shown that there is no continuous ice wall encircling the earth. Instead, Antarctica is a vast landmass covered in ice, with no distinct boundary separating it from the rest of the world.

Furthermore, satellite imagery and aerial surveys have provided a comprehensive view of the Antarctic continent, revealing its true shape and disproving the notion of an ice wall. These images clearly show a curved landmass, consistent with the round earth model.

In conclusion, the exploration of the Antarctic region and the scientific research conducted there have played a crucial role in debunking the flat earth theory. Early expeditions, modern scientific discoveries, and international cooperation have all contributed to our understanding of the round earth. The absence of an ice wall and the presence of a curved landmass in Antarctica provide compelling evidence against the flat earth theory. By embracing the reality of our round earth, we can move forward with a deeper appreciation for the wonders of our planet and the importance of scientific literacy.

3.3 Scientific Research and Discoveries

Scientific research and discoveries have played a crucial role in debunking the flat earth theory and uncovering the truth behind the supposed ice wall. Over the years, numerous scientific studies, experiments, and observations have provided overwhelming evidence in support of a round earth, while simultaneously refuting the claims made by flat earthers. In this section, we will explore some of the key scientific research and discoveries that have contributed to our understanding of the true shape of our planet and the absence of an ice wall.

3.3.1 Geodetic Surveys and Measurements

Geodetic surveys and measurements have been instrumental in determining the shape of the earth. These surveys involve the precise measurement of the Earth's surface and the calculation of its curvature. One of the most famous geodetic surveys was conducted by the French mathematician and astronomer Pierre Louis Maupertuis in the 18th century. Maupertuis led an expedition to Lapland, where his team measured the length of a degree of latitude. The results of this survey provided strong evidence for a round earth.

Since then, numerous geodetic surveys have been conducted all around the world, using advanced technologies and techniques. These surveys involve measuring the angles and distances between various points on the Earth's surface to create accurate models of its shape. The data collected from these surveys consistently supports the conclusion that the Earth is a sphere, with a slight bulge at the equator due to its rotation.

3.3.2 Satellite Imagery and Space Exploration

The advent of satellite imagery and space exploration has revolutionized our understanding of the Earth's shape and the absence of an ice wall. Satellites orbiting the Earth capture high-resolution images of our planet from different angles and perspectives. These images clearly show a spherical Earth, with no visible ice wall encircling it.

Furthermore, space exploration missions, such as those conducted by NASA and other space agencies, have provided irrefutable evidence of a round earth. Astronauts who have traveled to space have witnessed the curvature of the Earth firsthand and have captured stunning photographs and videos that clearly depict a spherical planet. These images and videos have been widely shared and have helped dispel the notion of a flat earth.

3.3.3 Oceanographic Research

Oceanographic research has also contributed to our understanding of the Earth's shape and the absence of an ice wall. Scientists studying ocean currents, tides, and the behavior of waves have gathered substantial evidence that supports a round earth. The way ocean currents flow and the behavior of waves can only be explained by a spherical Earth. If the Earth were flat, the patterns and behaviors observed in the oceans would be vastly different.

Additionally, the study of sea levels and the effects of gravity on the oceans further confirm the roundness of the Earth. The gravitational pull of a spherical Earth causes the oceans to bulge slightly at the equator, creating a measurable difference in sea level between the equator and the poles. This phenomenon, known as the geoid, is consistent with the shape of a round earth and is incompatible with the flat earth theory.

3.3.4 GPS and Global Communication Systems

The Global Positioning System (GPS) and other global communication systems rely on the assumption of a round earth to function accurately. GPS satellites orbiting the Earth transmit signals that are received by GPS receivers on the ground. These signals contain precise timing information, which allows the receivers to calculate their exact location on the Earth's surface.

The accuracy and reliability of GPS systems depend on the assumption that the Earth is a sphere. If the Earth were flat, GPS calculations would yield inaccurate results, rendering the system ineffective. The fact that GPS systems work flawlessly all around the world is a testament to the roundness of the Earth and the falsity of the flat earth theory.

In conclusion, scientific research and discoveries have overwhelmingly disproven the flat earth theory and revealed the truth about the absence of an ice wall. Geodetic surveys, satellite imagery, space exploration, oceanographic research, and the functioning of GPS systems all provide compelling evidence in support of a round earth. The accumulation of scientific knowledge and advancements in technology have allowed us to confidently embrace the reality of our round earth and dismiss the unfounded claims of the flat earth movement.

3.4 Debunking Claims about the Ice Wall

The concept of an ice wall surrounding the Earth is a central belief among flat Earthers. According to their claims, this ice wall acts as a barrier, preventing people from falling off the edge of the flat Earth. However, when we examine the evidence and scientific knowledge, we find that these claims are unfounded and can be easily debunked.

3.4.1 Lack of Physical Evidence

One of the main arguments put forth by flat Earthers is the existence of a massive ice wall encircling the Earth. They claim that this wall is thousands of miles long and several hundred feet high. However, despite these assertions, there is a distinct lack of physical evidence to support such claims.

No credible photographs or videos have ever been produced that clearly show the existence of this ice wall. In the age of modern technology and widespread access to cameras and satellites, it is highly unlikely that such a massive structure could remain hidden from public view. The absence of concrete evidence raises serious doubts about the existence of the ice wall.

3.4.2 Satellite Imagery and Exploration

Advancements in satellite technology have provided us with a wealth of information about our planet. Satellites orbiting the Earth capture high-resolution images of various regions, including the polar regions. These images clearly show vast expanses of ice and snow, but they do not depict an impenetrable ice wall encircling the Earth.

Furthermore, numerous scientific expeditions have been conducted to the polar regions, particularly Antarctica. These expeditions have involved researchers from different countries and various scientific disciplines. They have explored and mapped large portions of the Antarctic continent, but no evidence of an ice wall has been found.

If an ice wall truly existed, it would have been discovered and documented by these scientific expeditions. The absence of any credible scientific evidence supporting the existence of an ice wall further discredits the claims made by flat Earthers.

3.4.3 Circumnavigation and Flight Routes

Another compelling argument against the existence of an ice wall is the fact that people have successfully circumnavigated the Earth. Countless sailors, explorers, and aviators have traveled around the globe, crossing both the northern and southern hemispheres. If an ice wall were present, it would pose a significant obstacle to these journeys.

In reality, circumnavigation is possible because the Earth is a sphere. Flight routes, both commercial and private, are planned based on the understanding that the Earth is round. These flight paths take into account the curvature of the Earth and do not encounter any massive ice walls along the way.

The ability to circumnavigate the Earth and the absence of any reported encounters with an ice wall during these journeys strongly contradict the claims made by flat Earthers.

3.4.4 International Cooperation and Treaties

The Antarctic Treaty System, which includes the Antarctic Treaty and several related agreements, is an international agreement that governs the management and protection of Antarctica. This treaty has been signed by numerous countries, including those with differing political ideologies and agendas.

The existence of this treaty and the cooperation among nations in the exploration and scientific research of Antarctica further discredits the notion of an ice wall. If an ice wall were present, it would be impossible to maintain such a treaty without the knowledge and cooperation of all participating countries.

The fact that countries with diverse interests and ideologies have come together to protect and study Antarctica demonstrates the global consensus on the shape of our planet and the absence of an ice wall.

3.4.5 Logical Inconsistencies

When examining the claims made by flat Earthers regarding the ice wall, we encounter several logical inconsistencies. For instance, if the ice wall were truly impenetrable and encircled the Earth, how do ships and aircraft navigate through it? How do animals migrate across it? These questions remain unanswered by flat Earthers.

Additionally, the concept of an ice wall raises questions about the behavior of ocean currents and weather patterns. If the ice wall were present, it would significantly impact the flow of ocean currents and alter weather patterns around the globe. However, the observed behavior of ocean currents and weather systems aligns with the understanding of a round Earth, further undermining the claims of an ice wall.

In conclusion, the claims made by flat Earthers regarding the existence of an ice wall surrounding the Earth are not supported by scientific evidence or logical reasoning. The lack of physical evidence, the availability of satellite imagery, the successful circumnavigation of the Earth, international cooperation, and logical inconsistencies all contribute to the debunking of these claims. It is crucial to rely on scientific knowledge and evidence to understand the true nature of our planet.

4 Gravity and the Round Earth

4.1 Understanding Gravity and its Effects

Gravity is a fundamental force that plays a crucial role in shaping the Earth and the universe as a whole. In this section, we will explore the concept of gravity, its effects on our planet, and how it supports the evidence for a round Earth.

The Force of Gravity

Gravity is the force that attracts objects with mass towards each other. It is responsible for keeping our feet firmly planted on the ground and the planets in their orbits around the sun. Sir Isaac Newton first described gravity in his famous law of universal gravitation, which states that every object in the universe attracts every other object with a force that is directly proportional to their masses and inversely proportional to the square of the distance between them.

Gravity and Earth's Shape

One of the key pieces of evidence for a round Earth is the way gravity acts on the planet. Gravity pulls objects towards the center of mass, causing them to be evenly distributed around the Earth's surface. This distribution creates a spherical shape, as any excess mass would be pulled towards the center, resulting in a more compact shape.

If the Earth were flat, gravity would act differently. Objects would be pulled towards the center of the flat surface, causing an uneven distribution of mass. This would result in a bulging effect at the center and a lack of gravitational pull towards the edges. However, we observe that gravity acts uniformly in all directions, indicating a spherical shape.

Experiments and Observations on Gravity

Scientists have conducted numerous experiments and observations to study the effects of gravity and confirm the roundness of the Earth. One of the most famous experiments is the Cavendish experiment, performed by Henry Cavendish in 1798. This experiment measured the gravitational attraction between two small lead spheres and larger lead spheres, providing direct evidence of the existence of gravity.

Another experiment that supports the round Earth is the phenomenon of gravity anomalies. Gravity anomalies occur when there are variations in the gravitational field strength across the Earth's surface. These variations can be measured using specialized instruments, such as gravimeters, and are caused by differences in the density and distribution of mass within the Earth. The presence of gravity anomalies confirms the spherical shape of the Earth, as a flat Earth would not exhibit such variations.

Gravity's Influence on Earth's Atmosphere

Gravity not only shapes the Earth but also plays a vital role in maintaining our atmosphere. The Earth's gravitational pull keeps the atmosphere close to the surface, preventing it from escaping into space. Without gravity, the atmosphere would disperse, making it impossible for life as we know it to exist.

Additionally, gravity affects the behavior of gases in the atmosphere. It causes air to be denser near the surface and gradually decreases in density as we move higher into the atmosphere. This variation in density creates atmospheric pressure, which is essential for weather patterns, air circulation, and the overall stability of our planet's climate.

Conclusion

Understanding gravity and its effects is crucial in debunking the flat Earth theory. The force of gravity acts uniformly in all directions, resulting in a spherical shape for our planet. Experiments and observations have provided concrete evidence for the existence of gravity and its influence on Earth's shape and atmosphere. The concept of gravity aligns with the scientific consensus that the Earth is round, further discrediting the claims of the flat Earth theory.

4.2 Gravity's Role in Earth's Shape

Gravity is a fundamental force that plays a crucial role in shaping the Earth. It is the force that attracts objects towards the center of the planet, creating a spherical shape. In this section, we will explore the influence of gravity on Earth's shape and how it debunks the flat earth theory.

4.2.1 The Force of Gravity

Gravity is a force of attraction that exists between any two objects with mass. It is responsible for keeping our feet firmly planted on the ground and the planets in their orbits around the sun. The force of gravity is directly proportional to the mass of the objects and inversely proportional to the square of the distance between them.

On Earth, gravity pulls everything towards its center. This force acts uniformly in all directions, causing objects to be pulled towards the center of the planet. As a result, the Earth takes on a spherical shape, with the force of gravity pulling the mass towards the center, creating a balanced distribution.

4.2.2 The Equilibrium of Gravity

Gravity creates a state of equilibrium within the Earth's structure. The force of gravity acts equally on all parts of the planet, causing the material to be compressed towards the center. This compression results in a balanced distribution of mass, which ultimately leads to the spherical shape of the Earth.

If the Earth were flat, gravity would not be able to create this equilibrium. The force of gravity would cause the material to accumulate towards the center, resulting in a bulging effect at the center and a concave shape towards the edges. However, this is not observed in reality, as the Earth maintains a consistent spherical shape.

4.2.3 Gravity and the Earth's Surface

Gravity not only influences the overall shape of the Earth but also affects the surface of the planet. The force of gravity pulls objects towards the center, causing them to conform to the curvature of the Earth. This is evident in the way water behaves on the planet.

Water naturally seeks its lowest level due to the force of gravity. On a flat surface, water would spread out evenly, creating a perfectly flat surface. However, on a spherical Earth, gravity causes water to gather and form bodies of water such as oceans, seas, and lakes. The curvature of the Earth is also visible in the way ships disappear over the horizon, with only the top of their masts visible due to the Earth's curvature.

4.2.4 Gravity and Earth's Measurements

The influence of gravity on Earth's shape can be observed through various measurements and calculations. Geodetic surveys and measurements provide evidence of the Earth's curvature and its relationship with gravity. These surveys involve measuring the angles and distances between points on the Earth's surface to determine its shape.

By using precise instruments and mathematical calculations, geodetic surveys have consistently shown that the Earth is a sphere. The measurements taken across different locations on the planet align with the expected curvature based on the force of gravity. This further supports the scientific consensus that the Earth is round.

4.2.5 Gravity and Earth's Atmosphere

Gravity also plays a significant role in shaping and maintaining Earth's atmosphere. The force of gravity pulls the gases in the atmosphere towards the surface, creating a layer of air that surrounds the planet. Without gravity, the atmosphere would disperse into space, leaving the Earth devoid of the protective layer that sustains life.

The Earth's gravity holds the atmosphere close to its surface, preventing it from escaping into space. This gravitational force also creates a pressure gradient within the atmosphere, with higher pressure at lower altitudes and lower pressure at higher altitudes. This pressure gradient is essential for the circulation of air, the formation of weather patterns, and the regulation of temperature on Earth.

4.2.6 Gravity and the Flat Earth Theory

The concept of gravity directly contradicts the flat earth theory. If the Earth were flat, gravity would not be able to create a balanced distribution of mass, resulting in an uneven surface. The force of gravity would cause objects to accumulate towards the center, leading to a concave shape rather than a flat surface.

Furthermore, the behavior of water and the measurements taken across the Earth's surface consistently support the spherical shape of the planet. The curvature observed in everyday life, such as the disappearance of ships over the horizon, also aligns with the predictions based on the force of gravity.

In conclusion, gravity's role in shaping the Earth is undeniable. The force of gravity creates a state of equilibrium, pulling objects towards the center and resulting in a spherical shape. The evidence provided by geodetic surveys, the behavior of water, and the observations of Earth's surface all support the round earth model, debunking the flat earth theory. Gravity is a fundamental force that not only shapes our planet but also influences the behavior of our atmosphere, further solidifying the reality of our round Earth.

4.3 Experiments and Observations on Gravity

Gravity is a fundamental force that plays a crucial role in shaping the Earth and its atmosphere. Through a series of experiments and observations, scientists have been able to gather substantial evidence supporting the existence of gravity and its effects on our round Earth. In this section, we will explore some of these experiments and observations that debunk the flat Earth theory and reveal the reality of our round planet.

4.3.1 The Cavendish Experiment

One of the most famous experiments conducted to demonstrate the existence of gravity was the Cavendish experiment, performed by the British scientist Henry Cavendish in 1797-1798. This experiment involved measuring the gravitational attraction between two small lead spheres using a torsion balance.

Cavendish's experiment provided direct evidence of the gravitational force between objects and allowed for the determination of the gravitational constant. The results of this experiment confirmed the existence of gravity and its ability to act over large distances, supporting the round Earth model.

4.3.2 Falling Objects and Acceleration

Another compelling piece of evidence for gravity comes from the observation of falling objects. When objects are dropped from a height, they accelerate towards the Earth at a constant rate. This acceleration is known as the acceleration due to gravity and is approximately 9.8 meters per second squared.

The consistent acceleration of falling objects can be explained by the gravitational force exerted by the Earth. If the Earth were flat, we would expect objects to fall at different rates depending on their distance from the center of the Earth. However, the fact that all objects experience the same acceleration regardless of their mass or size strongly supports the idea of a spherical Earth with a uniform gravitational field.

4.3.3 Gravitational Anomalies

Gravity is not uniform across the Earth's surface due to variations in the distribution of mass. These variations result in gravitational anomalies, which can be measured using specialized instruments such as gravimeters. By mapping these anomalies, scientists can gain insights into the Earth's internal structure and composition.

The existence of gravitational anomalies is inconsistent with the flat Earth theory. If the Earth were flat, we would expect a uniform gravitational field across its surface. However, the presence of gravitational anomalies provides further evidence for the round Earth model, as it indicates variations in mass distribution beneath the Earth's surface.

4.3.4 Tides and the Moon's Influence

The phenomenon of tides is another compelling piece of evidence for the round Earth and the existence of gravity. Tides are the result of the gravitational interaction between the Earth, the Moon, and the Sun. The gravitational pull of the Moon causes the ocean waters to bulge, creating high tides, while the areas between the bulges experience low tides.

The predictable pattern of tides and their correlation with the positions of the Moon and the Sun strongly support the round Earth model. If the Earth were flat, it would be challenging to explain the consistent and predictable nature of tides across different locations on Earth.

4.3.5 Gravitational Lensing

Gravitational lensing is a phenomenon predicted by Einstein's theory of general relativity and provides further evidence for the existence of gravity.

When light passes near a massive object, such as a galaxy or a black hole, its path is bent due to the gravitational pull of the object. This bending of light can be observed and measured, confirming the presence of gravity.

The phenomenon of gravitational lensing has been observed and studied extensively, providing strong evidence for the round Earth model. If the Earth were flat, we would not expect to observe such gravitational lensing effects on light passing through space.

4.3.6 Satellite Orbits

The existence and behavior of satellites also provide evidence for the round Earth and the presence of gravity. Satellites orbit the Earth due to the gravitational attraction between the Earth and the satellite. The specific orbits and trajectories of satellites can be precisely calculated and predicted based on the laws of gravity.

If the Earth were flat, it would be impossible to explain the consistent and predictable orbits of satellites. The fact that satellites can maintain stable orbits around the Earth is a clear indication of the Earth's spherical shape and the presence of gravity.

In conclusion, a multitude of experiments and observations provide overwhelming evidence for the existence of gravity and the round Earth. The Cavendish experiment, the consistent acceleration of falling objects, gravitational anomalies, tides, gravitational lensing, and satellite orbits all support the round Earth model. These experiments and observations debunk the flat Earth theory and reveal the reality of our round planet.

4.4 Gravity's Influence on Earth's Atmosphere

Gravity, the fundamental force that governs the interactions between objects with mass, plays a crucial role in shaping and maintaining Earth's atmosphere. The atmosphere, a thin layer of gases surrounding our planet, is held in place by the gravitational pull of Earth. In this section, we will explore how gravity influences the composition, structure, and behavior of Earth's atmosphere.

4.4.1 Atmospheric Layers

Earth's atmosphere is divided into several distinct layers, each with its own unique characteristics. These layers are primarily defined by changes in temperature and composition, which are influenced by gravity. The layers include the troposphere, stratosphere, mesosphere, thermosphere, and exosphere.

The troposphere, the layer closest to the Earth's surface, is where weather phenomena occur. It extends from the surface up to an average altitude of about 12 kilometers (7.5 miles) at the poles and 18 kilometers (11 miles) at the equator. Gravity plays a significant role in maintaining the troposphere by keeping the air molecules close to the Earth's surface.

Above the troposphere lies the stratosphere, which extends from the top of the troposphere to an altitude of about 50 kilometers (31 miles). In the stratosphere, the concentration of ozone molecules increases, forming the ozone layer. This layer absorbs a significant amount of the Sun's ultraviolet radiation, protecting life on Earth. Gravity helps to confine the ozone layer within the stratosphere.

The mesosphere, located above the stratosphere, extends up to an altitude of about 85 kilometers (53 miles). In this layer, the temperature decreases with increasing altitude. Gravity plays a role in maintaining the mesosphere's stability and preventing the gases from escaping into space.

Beyond the mesosphere is the thermosphere, which extends up to an altitude of about 600 kilometers (370 miles). The thermosphere is characterized by high temperatures due to the absorption of intense solar radiation. Gravity helps to hold the thermosphere's gases in place, despite their high kinetic energy.

The outermost layer of Earth's atmosphere is the exosphere, which gradually transitions into the vacuum of space. The exosphere is composed of extremely low-density gases, such as hydrogen and helium, which can escape Earth's gravitational pull and disperse into space.

4.4.2 Atmospheric Pressure

Gravity also influences atmospheric pressure, which is the force exerted by the weight of the atmosphere on a unit area of the Earth's surface. As we move closer to the Earth's surface, the weight of the overlying air increases, resulting in higher atmospheric pressure. Conversely, as we move higher in the atmosphere, the weight of the air decreases, leading to lower atmospheric pressure.

The variation in atmospheric pressure with altitude has important implications for weather patterns and the distribution of gases in the atmosphere. It is the pressure gradient created by gravity that drives the movement of air masses, causing winds to blow from areas of high pressure to areas of low pressure.

4.4.3 Atmospheric Composition

Gravity also plays a role in determining the composition of Earth's atmosphere. The atmosphere is primarily composed of nitrogen (about 78%) and oxygen (about 21%), with trace amounts of other gases such as carbon dioxide, water vapor, and noble gases. These gases are held close to the Earth's surface by gravity, preventing them from escaping into space.

The gravitational force on Earth is strong enough to retain the lighter gases, such as hydrogen and helium, which have a tendency to escape into space. This is evident when we look at other celestial bodies, such as the Moon or Mars, which have much thinner atmospheres due to their weaker gravitational forces.

4.4.4 Atmospheric Circulation

Gravity also influences the circulation of the atmosphere, driving large-scale atmospheric motions such as global wind patterns and ocean currents. The differential heating of the Earth's surface by the Sun creates temperature and pressure gradients, which in turn generate atmospheric circulation.

The rotation of the Earth also plays a role in atmospheric circulation through the Coriolis effect, which causes moving air masses to be deflected to the right in the Northern Hemisphere and to the left in the Southern Hemisphere. This deflection, combined with gravity, helps to create the complex patterns of atmospheric circulation that we observe, including the trade winds, jet streams, and monsoons.

4.4.5 Atmospheric Escape

While gravity holds the majority of Earth's atmosphere in place, some gases can still escape into space. This process, known as atmospheric escape, occurs when the kinetic energy of gas molecules exceeds the gravitational pull of Earth. However, the rate of atmospheric escape is relatively slow, and Earth's atmosphere has been able to maintain its composition over billions of years.

In conclusion, gravity plays a fundamental role in shaping and maintaining Earth's atmosphere. It influences the structure, composition, and behavior of the atmosphere, from the formation of distinct layers to the circulation of air masses. Understanding the influence of gravity on Earth's atmosphere is crucial for comprehending the complex dynamics of our planet's climate and weather systems.

5 The Coriolis Effect and Earth's Rotation

5.1 Explaining the Coriolis Effect

The Coriolis effect is a phenomenon that occurs due to the rotation of the Earth. It is an important concept in understanding various natural phenomena, including weather patterns and projectile motion. In this section, we will delve into the details of the Coriolis effect and explore its implications on Earth's rotation.

5.1.1 Understanding the Coriolis Effect

The Coriolis effect is a result of the Earth's rotation on its axis. As the Earth spins, objects moving across its surface appear to be deflected from their straight paths. This deflection is caused by the difference in rotational speed between different latitudes. The Coriolis effect is responsible for the rotation of weather systems, the formation of ocean currents, and the trajectory of projectiles.

To understand the Coriolis effect, imagine standing at the North Pole and throwing a ball towards the equator. From your perspective, the ball would appear to curve to the right. Similarly, if you were standing at the equator and threw a ball towards the North Pole, it would appear to curve to the left. This apparent deflection is due to the Earth's rotation underneath the moving object.

5.1.2 Coriolis Effect and Weather Patterns

The Coriolis effect plays a crucial role in shaping global weather patterns. As air moves from high-pressure areas to low-pressure areas, it is deflected by the Coriolis effect. In the Northern Hemisphere, the deflection is to the right, while in the Southern Hemisphere, it is to the left.

This deflection causes the formation of large-scale wind systems, such as the trade winds and the prevailing westerlies. The Coriolis effect also influences the rotation of cyclones and anticyclones, determining their direction of rotation. Without the Coriolis effect, weather patterns would be drastically different, leading to significant changes in climate and atmospheric circulation.

5.1.3 Coriolis Effect and Projectile Motion

The Coriolis effect also affects the trajectory of moving objects, including projectiles. When an object is launched in a rotating reference frame, such as the Earth, it experiences a deflection due to the Coriolis effect. This deflection is perpendicular to the object's velocity and the axis of rotation.

For example, consider a long-range artillery shell fired towards a target. Due to the Earth's rotation, the shell experiences a deflection to the right in the Northern Hemisphere and to the left in the Southern Hemisphere. This deflection must be taken into account when aiming at distant targets, especially over long distances.

5.1.4 Observations and Experiments on Earth's Rotation

The Coriolis effect has been observed and measured in various experiments and observations. One notable experiment is the Foucault pendulum, which demonstrates the rotation of the Earth. The pendulum's swing appears to rotate slowly over time due to the Coriolis effect, providing evidence of Earth's rotation.

Another observation is the movement of hurricanes and cyclones. These weather systems rotate in a specific direction, which is consistent with the Coriolis effect. The deflection of ocean currents and the rotation of large-scale weather systems further confirm the existence of the Coriolis effect and Earth's rotation.

In conclusion, the Coriolis effect is a fundamental concept that explains the apparent deflection of moving objects on Earth's surface due to its rotation. It plays a crucial role in shaping weather patterns, determining the trajectory of projectiles, and influencing various natural phenomena. The Coriolis effect has been observed and measured through experiments and observations, providing concrete evidence of Earth's rotation. Understanding the Coriolis effect is essential in debunking the flat earth theory, as it directly contradicts the notion of a stationary, flat Earth.

5.2 Coriolis Effect and Weather Patterns

The Coriolis effect is a phenomenon that occurs due to the rotation of the Earth. It plays a significant role in shaping weather patterns and determining the direction of moving objects on our planet. Understanding the Coriolis effect is crucial in debunking the flat earth theory and revealing the reality of our round Earth.

5.2.1 The Coriolis Effect Explained

The Coriolis effect is a result of the Earth's rotation on its axis. As the Earth spins, objects moving across its surface appear to be deflected from their straight path. This deflection is observed in both the Northern and Southern Hemispheres, but in opposite directions. In the Northern Hemisphere, moving objects are deflected to the right, while in the Southern Hemisphere, they are deflected to the left.

The Coriolis effect is caused by the difference in rotational speed between different latitudes. Near the equator, the rotational speed is higher due to the larger circumference of the Earth. As one moves towards the poles, the rotational speed decreases. This difference in speed causes the deflection of moving objects.

5.2.2 Coriolis Effect and Weather Patterns

The Coriolis effect has a profound impact on weather patterns around the globe. It influences the direction of winds, the formation of cyclones and anticyclones, and the movement of ocean currents. Understanding these weather patterns and their correlation with the Coriolis effect provides strong evidence for the roundness of our planet.

In the Northern Hemisphere, the Coriolis effect causes winds to deflect to the right. This deflection leads to the formation of high-pressure systems, known as anticyclones, which rotate clockwise. On the other hand, low-pressure systems, or cyclones, form counterclockwise due to the deflection of winds to the right. These weather patterns are consistent with the rotation of a spherical Earth.

In the Southern Hemisphere, the Coriolis effect causes winds to deflect to the left. As a result, anticyclones rotate counterclockwise, while cyclones rotate clockwise. These weather patterns mirror those observed in the Northern Hemisphere, providing further evidence for the roundness of our planet.

5.2.3 Coriolis Effect and Projectile Motion

The Coriolis effect also affects the trajectory of moving objects, including projectiles. When an object is launched in a straight line on the Earth's surface, it appears to curve due to the Coriolis effect. In the Northern Hemisphere, the object curves to the right, while in the Southern Hemisphere, it curves to the left.

This phenomenon is particularly evident in long-range artillery fire and ballistic missile trajectories. Military forces around the world take the Coriolis effect into account when aiming their weapons, as failing to do so would result in significant inaccuracies. The consistent observation of these curved trajectories aligns with the spherical shape of our planet.

5.2.4 Observations and Experiments on Earth's Rotation

Numerous observations and experiments have been conducted to validate the Coriolis effect and, consequently, the Earth's rotation. One such experiment is the Foucault pendulum, which demonstrates the rotation of the Earth by the changing direction of the pendulum's swing.

The Foucault pendulum consists of a long, heavy wire with a weight attached at the end. When set in motion, the pendulum's swing appears to rotate slowly in a clockwise direction in the Northern Hemisphere and counterclockwise in the Southern Hemisphere. This rotation is a direct result of the Coriolis effect and provides tangible evidence of the Earth's rotation.

Furthermore, satellite imagery and global positioning systems (GPS) rely on the Earth's rotation and the Coriolis effect to accurately determine positions on the planet's surface. These technologies would not function as they do if the Earth were flat. The consistent and reliable data obtained from these systems further supports the reality of our round Earth.

In conclusion, the Coriolis effect is a fundamental aspect of our planet's rotation and plays a crucial role in shaping weather patterns and the trajectory of moving objects. The observations and experiments conducted on the Coriolis effect provide compelling evidence for the roundness of the Earth. By understanding and embracing the Coriolis effect, we can debunk the flat earth theory and reveal the undeniable reality of our round Earth.

5.3 Coriolis Effect and Projectile Motion

The Coriolis effect is a phenomenon that occurs due to the rotation of the Earth. It has a significant impact on various aspects of our daily lives, including weather patterns, ocean currents, and even the trajectory of projectiles. Understanding the Coriolis effect is crucial in debunking the flat earth theory and revealing the reality of our round Earth.

5.3.1 Explaining the Coriolis Effect

The Coriolis effect is a result of the Earth's rotation causing moving objects, such as air or water, to be deflected from their straight path. This deflection occurs because different points on the Earth's surface are moving at different speeds due to the Earth's rotation. As a result, objects moving across the Earth's surface appear to curve from their intended path.

To understand the Coriolis effect, imagine standing at the North Pole and throwing a ball towards the equator. From your perspective, the ball would appear to curve to the right. Similarly, if you were standing at the equator and threw a ball towards the North Pole, it would appear to curve to the left. This deflection is caused by the difference in rotational speed between the two points.

5.3.2 Coriolis Effect and Weather Patterns

The Coriolis effect plays a crucial role in shaping global weather patterns. It influences the direction of winds and the formation of cyclones and anticyclones. In the Northern Hemisphere, the Coriolis effect causes winds to deflect to the right, while in the Southern Hemisphere, it causes winds to deflect to the left.

For example, the trade winds, which blow from east to west in the tropics, are a result of the Coriolis effect. As warm air rises near the equator and moves towards the poles, it is deflected by the Earth's rotation, creating the trade winds. These winds have a significant impact on global weather patterns and are evidence of the Earth's rotation.

5.3.3 Coriolis Effect and Projectile Motion

The Coriolis effect also affects the trajectory of projectiles, such as missiles, bullets, or even thrown objects. When an object is launched from a moving platform, such as a rotating Earth, its path is influenced by the Coriolis effect.

In the Northern Hemisphere, projectiles traveling long distances tend to be deflected to the right, while in the Southern Hemisphere, they are deflected to the left. This deflection is due to the difference in rotational speed between the launch point and the target.

For example, consider a long-range missile launched from the Northern Hemisphere towards the equator. Due to the Coriolis effect, the missile would experience a deflection to the right during its flight. This deflection needs to be taken into account when calculating the trajectory of the missile to ensure it reaches its intended target.

5.3.4 Observations and Experiments on Earth's Rotation

Numerous observations and experiments have been conducted to confirm the Earth's rotation and the existence of the Coriolis effect. One of the most famous experiments is the Foucault pendulum, invented by French physicist Léon Foucault in 1851.

The Foucault pendulum consists of a long, heavy pendulum suspended from a fixed point. As the pendulum swings back and forth, its plane of oscillation appears to rotate due to the Earth's rotation. This rotation is a direct result of the Coriolis effect and provides visual evidence of the Earth's rotation.

In addition to the Foucault pendulum, other experiments, such as the Coriolis effect on rotating platforms and the behavior of hurricanes in different hemispheres, further support the existence of the Coriolis effect and the Earth's rotation.

By understanding and observing the Coriolis effect, we can confidently conclude that the Earth is round and rotating. The deflection of winds, the trajectory of projectiles, and the behavior of pendulums all provide tangible evidence of the Earth's rotation. These observations directly contradict the claims of the flat earth theory, further revealing the reality of our round Earth.

The Coriolis effect is just one piece of the puzzle in debunking the flat earth theory. In the next section, we will explore the concept of Earth's curvature and how it can be observed and measured in our everyday lives.

5.4 Observations and Experiments on Earth's Rotation

The rotation of the Earth is a fundamental aspect of our planet's existence, and it plays a crucial role in shaping our understanding of its roundness. Over the centuries, scientists and researchers have conducted numerous observations and experiments to study and confirm the Earth's rotation. These investigations have provided compelling evidence that supports the round Earth model and debunks the flat Earth theory. In this section, we will explore some of the key observations and experiments that have contributed to our understanding of Earth's rotation.

5.4.1 Foucault Pendulum

One of the most famous experiments demonstrating the Earth's rotation is the Foucault pendulum. In 1851, the French physicist Léon Foucault suspended a long pendulum from the ceiling of the Panthéon in Paris. As the pendulum swung back and forth, it appeared to change its direction of swing over time. This phenomenon, known as the precession of the Foucault pendulum, is caused by the rotation of the Earth.

The precession of the pendulum occurs because the Earth rotates underneath it while the pendulum maintains its plane of oscillation. This experiment provided direct visual evidence of the Earth's rotation and demonstrated that the Earth is not stationary but rather spinning on its axis.

5.4.2 Coriolis Effect

Another compelling piece of evidence for Earth's rotation is the Coriolis effect. The Coriolis effect is the apparent deflection of moving objects, such as winds or ocean currents, caused by the rotation of the Earth. This effect is responsible for the rotation of weather systems and the formation of cyclones and anticyclones.

The Coriolis effect can be observed in the movement of projectiles as well. For example, when a long-range artillery shell is fired, it experiences a slight deflection due to the Earth's rotation. This deflection is taken into account by artillery operators to ensure accurate targeting.

5.4.3 Star Trails

Observing the movement of stars in the night sky also provides evidence of Earth's rotation. When we look at the stars over an extended period, we can observe them appearing to move in circular paths around a fixed point. This phenomenon is known as star trails.

The circular paths of star trails are a result of the Earth's rotation on its axis. As the Earth spins, different stars become visible at different times, creating the illusion of their movement across the sky. The rotation of the Earth is responsible for the consistent pattern of star trails observed from different locations on the planet.

5.4.4 Time Zones

The existence of time zones is another piece of evidence that supports the Earth's rotation. Time zones are regions of the Earth that have the same standard time. They are based on the concept that as the Earth rotates, different parts of the planet experience daylight and darkness at different times.

If the Earth were flat and stationary, there would be no need for time zones. However, the fact that we have a system of time zones around the world, with each zone representing a specific hour offset from Coordinated Universal Time (UTC), confirms the Earth's rotation. As we move from one time zone to another, we are adjusting our clocks to account for the Earth's rotation and the changing position of the Sun in the sky.

5.4.5 Satellite Observations

Satellites orbiting the Earth provide further evidence of our planet's rotation. Satellites are launched into space with specific velocities and trajectories to ensure that they remain in stable orbits around the Earth. These orbits are calculated based on the Earth's rotation and gravitational forces.

If the Earth were flat, satellites would not be able to maintain their orbits as they do. The fact that satellites can successfully orbit the Earth and provide us with valuable data and communication services is a testament to the Earth's roundness and rotation.

5.4.6 Astronomical Phenomena

Various astronomical phenomena also support the concept of Earth's rotation. For instance, the phenomenon of the equinoxes, where the Sun appears to cross the celestial equator, is a direct result of the Earth's axial tilt and rotation. Similarly, the changing positions of constellations throughout the year can be explained by the Earth's rotation around the Sun.

Observations of other celestial bodies, such as the Moon and other planets, also provide evidence of Earth's rotation. The phases of the Moon, lunar eclipses, and the apparent retrograde motion of planets are all consistent with the Earth's rotation and its position in the solar system.

In conclusion, a wealth of observations and experiments have provided compelling evidence for the Earth's rotation. The Foucault pendulum, the Coriolis effect, star trails, time zones, satellite observations, and various astronomical phenomena all support the round Earth model. These observations and experiments have debunked the flat Earth theory and reinforced our understanding of the reality of our round Earth.

6 The Curvature of Earth

6.1 Curvature Calculations and Formulas

One of the most compelling pieces of evidence that supports the round Earth theory is the concept of curvature. The curvature of the Earth refers to the gradual bending of its surface, which is a direct result of its spherical shape. In this section, we will explore the calculations and formulas used to determine the curvature of the Earth and how they provide undeniable proof of its roundness.

6.1.1 The Earth's Radius

To understand the curvature of the Earth, we must first establish its radius. The radius of the Earth is the distance from its center to any point on its surface. Through centuries of scientific research and measurements, the average radius of the Earth has been determined to be approximately 6,371 kilometers (3,959 miles). This value serves as a fundamental parameter in curvature calculations.

6.1.2 Curvature Formula

The curvature of the Earth can be calculated using a simple formula that takes into account the radius of the Earth and the distance between two points on its surface. This formula is known as the "curvature formula" and is expressed as:

$$C = (d^2) / (2R)$$

Where: - C represents the curvature of the Earth - d is the distance between two points on the Earth's surface - R is the radius of the Earth

By plugging in the appropriate values for d and R, we can determine the curvature between any two points on the Earth's surface.

6.1.3 Example Calculation

Let's consider an example to illustrate the application of the curvature formula. Suppose we want to calculate the curvature between two cities that are 100 kilometers apart. Using the formula, we can calculate the curvature as follows:

$$C = (100^2) / (2 * 6{,}371)$$
$$C = 10{,}000 / 12{,}742$$
$$C \approx 0.785 \text{ kilometers}$$

This means that over a distance of 100 kilometers, the Earth's surface curves downward by approximately 0.785 kilometers. This curvature is not easily noticeable to the naked eye, but it becomes more apparent over longer distances.

6.1.4 The Horizon and Curvature

One of the most visible manifestations of Earth's curvature is the way it affects our perception of the horizon. As we stand on a flat surface, such as a beach or a plain, the horizon appears as a straight line where the sky meets the Earth. However, as we move higher above the surface, such as in an airplane or on a mountain, the curvature of the Earth becomes more apparent.

As we ascend, the horizon begins to curve, gradually revealing more of the Earth's surface. This phenomenon is a direct result of the Earth's spherical shape and the way light travels in straight lines. The curvature calculations and formulas we discussed earlier help explain why the horizon appears to curve as we gain altitude.

6.1.5 Observing Curvature in Everyday Life

While the curvature of the Earth may not be immediately noticeable in our day-to-day lives, there are several ways we can observe it if we know where to look. For example, when watching a ship sail away from the shore, it gradually disappears from view, starting with the hull and eventually the mast. This phenomenon, known as "the sinking ship effect," occurs because the ship is moving beyond the horizon, which is curved.

Similarly, when observing the sunset or sunrise, we can witness the gradual disappearance or appearance of the Sun's disk as it moves below or above the horizon. This phenomenon is also a result of the Earth's curvature, as the Sun's light is obstructed by the curvature of the Earth.

6.1.6 Curvature and Horizon Observations

The curvature of the Earth also has implications for long-distance observations and measurements. For instance, when looking at a distant object, such as a mountain range or a tall building, the bottom part of the object appears hidden or obscured by the Earth's curvature. As we move closer to the object, more of it becomes visible, eventually revealing its full height.

This observation aligns with the curvature calculations and formulas we discussed earlier. It provides further evidence that the Earth is not flat but rather curved, as objects beyond the horizon gradually become visible as we approach them.

6.1.7 Photographic Evidence of Earth's Curvature

In the age of photography and advanced imaging technology, we have access to countless images that capture the curvature of the Earth. Aerial photographs taken from high altitudes, such as those captured by astronauts aboard the International Space Station, clearly show the Earth's curved horizon.

Additionally, photographs taken from the surface of the Earth, such as those captured from mountaintops or tall buildings, also provide visual evidence of the Earth's curvature. These images consistently depict a curved horizon, further reinforcing the scientific understanding of our planet's shape.

In conclusion, the calculations and formulas used to determine the curvature of the Earth provide undeniable evidence of its roundness. Observations of the horizon, the sinking ship effect, and photographic evidence all support the scientific consensus that the Earth is not flat but rather a spherical body. Understanding the curvature of the Earth is crucial in debunking the flat Earth theory and embracing the reality of our round planet.

6.2 Visible Curvature in Everyday Life

One of the most compelling pieces of evidence that supports the round Earth theory is the visible curvature that can be observed in everyday life. While it may not be immediately apparent to the naked eye, there are several instances where the curvature of the Earth becomes evident, providing a clear indication of our planet's shape.

6.2.1 Observing the Horizon

One of the simplest ways to observe the curvature of the Earth is by looking at the horizon. When standing on a beach or a large open field, you can see that the horizon appears to curve slightly downward, creating a distinct arc. This curvature is a result of the Earth's spherical shape, as light from distant objects is gradually blocked by the curvature of the planet.

As you gaze out into the distance, you may notice that ships or boats gradually disappear from view as they sail away. This phenomenon, known as "sinking ship effect," occurs because the curvature of the Earth obstructs our line of sight. The bottom of the ship disappears first, followed by the rest of the vessel, until only the mast or the highest point remains visible. This gradual disappearance is consistent with the curvature of a spherical Earth.

6.2.2 Aerial Views and High Altitude Photography

Another way to observe the curvature of the Earth is through aerial views and high altitude photography. When flying in an airplane or looking at photographs taken from space, the curvature becomes even more apparent. As the altitude increases, the horizon appears to curve more prominently, providing a clear visual representation of the Earth's roundness.

Astronauts who have had the privilege of viewing Earth from space often describe the awe-inspiring sight of seeing our planet as a beautiful blue sphere suspended in the vastness of space. The curvature of the Earth is unmistakable from this vantage point, reinforcing the overwhelming evidence that our planet is indeed round.

6.2.3 The Drop in Temperature with Altitude

Another interesting phenomenon that supports the round Earth theory is the drop in temperature with increasing altitude. As you ascend to higher altitudes, such as when climbing a mountain, you may notice that the temperature gradually decreases. This temperature drop occurs because the atmosphere becomes thinner at higher altitudes, and the decrease in air pressure leads to a decrease in temperature.

This drop in temperature is consistent with the curvature of the Earth. As you climb higher, you are moving away from the Earth's surface and entering regions of the atmosphere that are farther from the planet's heat source, the sun. The curvature of the Earth causes this gradual decrease in temperature, providing further evidence of our planet's spherical shape.

6.2.4 The Coriolis Effect

The Coriolis effect is another phenomenon that provides evidence for the curvature of the Earth. The Coriolis effect is the apparent deflection of moving objects caused by the rotation of the Earth. It affects the movement of air masses, ocean currents, and even the flight paths of airplanes.

For example, in the Northern Hemisphere, the Coriolis effect causes moving objects to be deflected to the right. This deflection is a result of the Earth's rotation and the varying speeds of different latitudes. The Coriolis effect is consistent with the curvature of the Earth, as the deflection occurs due to the rotation of a spherical planet.

6.2.5 The Curvature of Large Structures

The curvature of the Earth is also evident in the construction of large structures such as bridges and railways. Engineers and architects must take into account the curvature of the Earth when designing these structures to ensure their stability and safety.

For instance, when constructing a long bridge, engineers must account for the curvature of the Earth to ensure that the bridge is level and does not sag in the middle. Similarly, when laying railway tracks over long distances, the curvature of the Earth must be considered to ensure that the tracks are properly aligned and do not deviate from a straight path.

These examples demonstrate that the curvature of the Earth is not just a theoretical concept but has practical implications in various fields of engineering and construction.

In conclusion, the visible curvature of the Earth in everyday life provides compelling evidence for the round Earth theory. Observing the curvature of the horizon, aerial views, the drop in temperature with altitude, the Coriolis effect, and the construction of large structures all point to the undeniable fact that our planet is indeed spherical. These observations, combined with the wealth of scientific evidence, leave no room for doubt about the true shape of our Earth.

6.3 Curvature and Horizon Observations

One of the most compelling pieces of evidence that supports the round Earth theory is the observation of curvature and the way it affects our view of the horizon. When we look out at the vast expanse of the ocean or stand on a high mountain, we can see the curvature of the Earth with our own eyes. This observation directly contradicts the claims made by flat Earth proponents and provides strong evidence for the reality of our round planet.

6.3.1 The Horizon and its Curvature

When we stand on a beach and gaze out at the ocean, we can see that the horizon appears to be a straight line where the sky meets the water. However, this is an optical illusion caused by the curvature of the Earth. In reality, the Earth's surface is curved, and the horizon is actually a circular line that extends around us.

As we move higher above the Earth's surface, such as climbing a mountain or flying in an airplane, the curvature becomes more apparent. The higher we go, the more we can see the Earth's curvature, and the horizon appears to dip downwards. This observation is consistent with the spherical shape of the Earth and is not possible on a flat surface.

6.3.2 The Dip of the Horizon

One of the most fascinating aspects of the curvature of the Earth is the way it affects our perception of distant objects. As we look out at the horizon, objects that are farther away appear to be partially hidden or "dipped" below the horizon line. This phenomenon is known as the dip of the horizon.

The amount of dip depends on the distance between the observer and the object. The farther away an object is, the more it appears to be hidden below the horizon. This is because the Earth's surface curves away from our line of sight, causing objects to gradually disappear from view as they move farther away.

For example, if we observe a ship sailing towards the horizon, we can see that it gradually disappears from view as it moves farther away. The bottom of the ship is the first to disappear, followed by the middle, and eventually the entire ship is hidden from sight. This observation is consistent with the curvature of the Earth and cannot be explained by a flat Earth model.

6.3.3 Calculating the Curvature

The curvature of the Earth can be calculated using mathematical formulas and measurements. One commonly used formula is the Pythagorean theorem, which relates the curvature of the Earth to the distance and height of the observer.

By measuring the height of an observer above sea level and the distance to the horizon, we can calculate the curvature of the Earth. These calculations consistently yield results that are consistent with a spherical Earth and provide further evidence against the flat Earth theory.

6.3.4 Horizon Observations and Photography

In addition to our direct observations of the curvature of the Earth, photography has played a crucial role in documenting and confirming the round shape of our planet. Over the years, countless photographs have been taken from various altitudes and locations, all of which consistently show the curvature of the Earth.

From photographs taken by astronauts in space to images captured by high-altitude balloons and even everyday photographs taken by travelers, the evidence for the curvature of the Earth is overwhelming. These photographs not only provide visual proof of the Earth's roundness but also serve as a powerful tool for educating and dispelling the misconceptions surrounding the flat Earth theory.

In conclusion, the observation of curvature and the way it affects our view of the horizon is a compelling piece of evidence that supports the reality of our round Earth. From the dip of the horizon to the calculations of curvature and the countless photographs taken from various altitudes, all the evidence points to a spherical Earth. These observations directly contradict the claims made by flat Earth proponents and provide a solid foundation for understanding the true shape of our planet.

6.4 Photographic Evidence of Earth's Curvature

One of the most compelling pieces of evidence that supports the round Earth theory is the abundance of photographic evidence showcasing the curvature of our planet. These photographs, taken from various vantage points, provide undeniable visual proof of Earth's spherical shape. In this section, we will explore some of the most iconic and influential photographs that have helped debunk the flat Earth theory.

6.4.1 The Blue Marble

One of the most famous photographs of Earth is the "Blue Marble" image taken by the Apollo 17 crew in 1972. This stunning photograph shows our planet as a beautiful, round orb suspended in the vastness of space. The Blue Marble image provides a clear view of Earth's curvature, with the continents and oceans clearly visible. This photograph, taken from a distance of about 29,000 kilometers (18,000 miles) away, is a powerful visual representation of our planet's true shape.

6.4.2 The Earthrise

Another iconic photograph that showcases Earth's curvature is the "Earthrise" image captured by the Apollo 8 crew in 1968. This photograph was taken during the first manned mission to orbit the Moon and shows Earth rising above the lunar horizon. The Earthrise image not only reveals the curvature of our planet but also highlights the fragility and beauty of our home in the vastness of space. This photograph had a profound impact on human consciousness, reminding us of the interconnectedness of all life on Earth.

6.4.3 The ISS Perspective

In more recent years, photographs taken from the International Space Station (ISS) have provided further evidence of Earth's curvature. Astronauts aboard the ISS regularly capture breathtaking images of our planet from their unique vantage point in low Earth orbit. These photographs, often shared on social media, offer a stunning perspective of Earth's roundness. The curvature of the planet is clearly visible in these images, reinforcing the overwhelming scientific consensus that Earth is indeed a sphere.

6.4.4 High-Altitude Balloon Photography

Photographic evidence of Earth's curvature is not limited to images taken from space. High-altitude balloon photography has also played a significant role in debunking the flat Earth theory. Amateur photographers and scientists alike have launched high-altitude balloons equipped with cameras to capture images of Earth from the edge of space. These photographs consistently reveal a distinct curvature, providing further confirmation of our planet's spherical shape.

6.4.5 Horizon Observations

In addition to photographs taken from space and high-altitude balloons, everyday observations of the horizon also provide evidence of Earth's curvature. When standing on a beach or atop a tall building, one can observe that as they look out into the distance, the horizon appears to curve downward. This observation aligns with the mathematical calculations and formulas that describe Earth's curvature. The fact that the horizon appears curved from various vantage points on Earth further supports the notion that our planet is round.

6.4.6 Panoramic Views

Panoramic photographs taken from elevated locations, such as mountaintops or tall buildings, also offer compelling evidence of Earth's curvature. When capturing a wide-angle view of the landscape, the curvature of the Earth becomes apparent as the horizon curves away from the observer. These panoramic images provide a visual representation of the spherical shape of our planet and serve as a powerful tool in debunking the flat Earth theory.

In conclusion, the abundance of photographic evidence showcasing Earth's curvature leaves little room for doubt regarding the shape of our planet. The Blue Marble, Earthrise, ISS photographs, high-altitude balloon images, horizon observations, and panoramic views all provide compelling visual proof of Earth's spherical shape. These photographs, taken from various vantage points and perspectives, consistently reveal a distinct curvature, reinforcing the overwhelming scientific consensus that our planet is round. The photographic evidence of Earth's curvature stands as a testament to the power of observation and scientific inquiry in uncovering the truth about our world.

7 The Round Earth in Space

7.1 Earth's Position in the Solar System

The Earth, our home planet, occupies a unique position in the vast expanse of the solar system. Understanding our position in relation to the other celestial bodies is crucial in unraveling the truth about the shape of our planet. Through centuries of scientific exploration and observation, we have gathered compelling evidence that supports the round Earth theory and debunks the notion of a flat Earth.

Earth's Orbit around the Sun

One of the fundamental aspects of our position in the solar system is Earth's orbit around the Sun. The Earth follows an elliptical path around the Sun, with the Sun at one of the foci of the ellipse. This orbit is not a perfect circle but rather a slightly elongated shape, resulting in variations in our distance from the Sun throughout the year. This elliptical orbit is responsible for the changing seasons and the length of our days.

The Sun's Influence on Earth

The Sun, as the central star of our solar system, plays a crucial role in shaping Earth's position and characteristics. Its gravitational force keeps the Earth in its orbit, preventing it from drifting away into space. The Sun's immense mass creates a gravitational pull that keeps all the planets, including Earth, in their respective orbits.

Furthermore, the Sun's energy is the primary source of heat and light for our planet. The energy radiated by the Sun warms the Earth's surface, drives weather patterns, and sustains life as we know it. The Sun's position in the solar system, along with its gravitational influence, is a testament to the round Earth theory.

Earth's Relationship with Other Planets

In addition to its orbit around the Sun, Earth also interacts with other planets in the solar system. The gravitational forces exerted by neighboring planets, such as Venus and Mars, have a minor influence on Earth's orbit. These interactions, known as planetary perturbations, cause slight variations in the shape and orientation of Earth's orbit over long periods.

Furthermore, the alignment of the planets and their positions in the sky can be accurately predicted using mathematical models based on the round Earth theory. These predictions have been consistently confirmed through observations and measurements, providing further evidence of Earth's position in the solar system.

The Moon and Earth's Tides

Another significant aspect of Earth's position in the solar system is its relationship with the Moon. The Moon, Earth's only natural satellite, exerts a gravitational force on our planet. This gravitational interaction between the Earth and the Moon is responsible for the tides we observe in our oceans.

The Moon's gravitational pull causes a bulge in the Earth's oceans, resulting in the rise and fall of tides. The predictable patterns of high and low tides across the globe can be accurately calculated based on the round Earth theory and the Moon's position in relation to our planet.

Planetary Observations and Space Missions

Throughout history, humans have sent numerous spacecraft and satellites into space to explore the solar system. These missions have provided us with invaluable data and images that confirm the round Earth theory. Spacecraft like Voyager, Cassini, and New Horizons have captured breathtaking images of Earth from different vantage points, showcasing its spherical shape.

Additionally, astronauts who have traveled to space have witnessed the curvature of the Earth firsthand. Their experiences and photographs from space missions, such as the iconic "Blue Marble" image taken during the Apollo missions, provide undeniable evidence of Earth's roundness.

Conclusion

The position of Earth in the solar system, its orbit around the Sun, and its interactions with other celestial bodies all point to the reality of our round Earth. The evidence gathered through centuries of scientific exploration, observations, and space missions leaves no room for doubt. The flat Earth theory is thoroughly debunked by the overwhelming evidence supporting the round Earth theory. Understanding our position in the solar system is crucial in embracing the reality of our round Earth and dispelling the misconceptions surrounding the flat Earth theory.

7.2 Orbits and Gravitational Forces

In this section, we will explore the concept of orbits and gravitational forces, which play a crucial role in understanding the reality of our round Earth. The understanding of orbits and gravitational forces has been instrumental in debunking the flat Earth theory and providing evidence for the spherical shape of our planet.

7.2.1 The Concept of Orbits

An orbit is the path followed by an object around a more massive celestial body, such as a planet or a star. The concept of orbits was first explained by Johannes Kepler in the 17th century, who formulated three laws of planetary motion. These laws describe the behavior of objects in space and provide a solid foundation for understanding the mechanics of orbits.

Kepler's first law, known as the law of elliptical orbits, states that planets and other celestial bodies move in elliptical paths around the Sun, with the Sun located at one of the foci of the ellipse. This law directly contradicts the flat Earth theory, which suggests a stationary Earth at the center of the universe.

7.2.2 Gravitational Forces

Gravitational forces are the driving force behind the formation of orbits. Sir Isaac Newton's law of universal gravitation, formulated in the 17th century, explains the force of attraction between two objects with mass. According to this law, every object in the universe attracts every other object with a force that is directly proportional to the product of their masses and inversely proportional to the square of the distance between them.

The gravitational force between the Earth and other celestial bodies, such as the Moon and artificial satellites, is what keeps them in orbit. This force acts as a centripetal force, constantly pulling the objects towards the Earth while simultaneously causing them to move in a curved path. If the Earth were flat, it would be impossible for objects to maintain a stable orbit around it.

7.2.3 Satellite Orbits

Satellites, both natural and artificial, provide compelling evidence for the round Earth. Natural satellites, such as the Moon, have been observed for centuries and are known to orbit the Earth. Artificial satellites, on the other hand, are man-made objects intentionally placed in orbit around the Earth for various purposes, including communication, weather monitoring, and scientific research.

The orbits of artificial satellites are carefully calculated and designed to ensure their stability and functionality. Satellites are launched into space using rockets, and once they reach the desired altitude, they are placed into specific orbits using onboard propulsion systems. These orbits can be circular or elliptical, depending on the mission requirements.

The fact that satellites can maintain stable orbits around the Earth is a clear indication that the Earth is not flat. If the Earth were flat, satellites would not be able to maintain a consistent altitude and would eventually fall out of orbit or crash into the Earth's surface.

7.2.4 The Role of Gravity in Orbits

Gravity plays a crucial role in determining the shape and stability of orbits. The gravitational force between the Earth and an object in orbit provides the necessary centripetal force to keep the object moving in a curved path. The strength of the gravitational force depends on the mass of the Earth and the distance between the Earth and the object.

Objects in low Earth orbit, such as the International Space Station (ISS), experience a slightly weaker gravitational force compared to objects on the Earth's surface. This is because the gravitational force decreases with distance from the Earth's center. However, the difference in gravitational force is not significant enough to cause objects in orbit to fall towards the Earth.

The ability of objects to remain in orbit around the Earth is a direct result of the balance between the gravitational force pulling them towards the Earth and their forward velocity. This delicate balance allows satellites and other objects to continuously fall towards the Earth while simultaneously moving forward, resulting in a stable orbit.

7.2.5 The Predictability of Orbits

The understanding of orbits and gravitational forces has allowed scientists to accurately predict the motion of celestial bodies and artificial satellites. By applying the laws of motion and the principles of gravity, scientists can calculate the trajectory and position of objects in space with remarkable precision.

The predictability of orbits is a testament to the accuracy of our understanding of the laws of physics and the shape of the Earth. If the Earth were flat, the calculations and predictions of satellite orbits would not align with the observed data. However, the consistency between the predicted and observed orbits provides strong evidence for the round Earth.

In conclusion, the concept of orbits and gravitational forces provides compelling evidence against the flat Earth theory. The ability of objects to maintain stable orbits around the Earth, the predictability of satellite trajectories, and the understanding of gravitational forces all point towards the spherical shape of our planet. The scientific understanding of orbits and gravitational forces has played a crucial role in debunking the flat Earth theory and revealing the reality of our round Earth.

7.3 Satellites and Communication Systems

Satellites play a crucial role in our modern world, enabling global communication, navigation, weather forecasting, and scientific research. They are an essential component of our understanding of the round Earth. In this section, we will explore how satellites work and how they provide evidence for the reality of our round Earth.

7.3.1 The Functioning of Satellites

Satellites are man-made objects that orbit around the Earth. They are launched into space and placed in specific orbits to perform various tasks. Communication satellites, for example, are designed to transmit and receive signals for television, telephone, and internet communication. Weather satellites provide valuable data for meteorological predictions, while navigation satellites, such as the Global Positioning System (GPS), allow us to determine our precise location on Earth.

Satellites operate by utilizing the principles of physics and engineering. They are equipped with antennas, sensors, and transmitters that enable them to send and receive signals. These signals are transmitted to and from ground-based stations, which then relay the information to the intended recipients. The ability of satellites to communicate with multiple ground stations across different continents is a testament to the global nature of our round Earth.

7.3.2 Satellite Orbits and Earth's Shape

The orbits of satellites are carefully planned to ensure their stability and functionality. Satellites can be placed in different types of orbits, including geostationary orbits and low Earth orbits. Geostationary satellites are positioned at a specific altitude above the equator, where they appear to remain stationary relative to a point on Earth's surface. This is possible because the satellite's orbital period matches the rotation period of the Earth.

The fact that geostationary satellites exist and function as intended is strong evidence for the round Earth. If the Earth were flat, it would be impossible for satellites to maintain a fixed position above a specific point on the planet's surface. The ability to establish and maintain geostationary orbits is a direct result of Earth's spherical shape.

7.3.3 Communication Systems and Satellite Signals

Satellites are equipped with sophisticated communication systems that allow them to transmit and receive signals over vast distances. These signals are transmitted using radio waves, which can travel through the atmosphere and space. The signals are then received by ground-based stations, which decode and process the information.

The reliability and accuracy of satellite communication systems further support the round Earth model. For example, GPS satellites provide precise location data by triangulating signals from multiple satellites. This would not be possible if the Earth were flat, as the distances and angles between satellites and receivers would not align with the expected measurements.

Furthermore, the ability to establish and maintain communication with satellites in different parts of the world simultaneously is only possible because the Earth is round. If the Earth were flat, communication signals would be limited to a specific range and direction, making global communication impossible.

7.3.4 Scientific Research and Satellite Observations

Satellites are invaluable tools for scientific research and observations. They provide a unique vantage point from which scientists can study Earth's atmosphere, climate patterns, ocean currents, and even the movement of tectonic plates. Satellites equipped with advanced sensors and instruments capture data that helps us understand the complex systems and processes occurring on our planet.

The data collected by satellites has provided significant evidence for the round Earth. For example, satellite imagery shows the curvature of Earth's surface, with the horizon appearing curved when viewed from space. These images, combined with other scientific measurements and observations, confirm the spherical shape of our planet.

Satellites also contribute to our understanding of Earth's rotation. By monitoring the movement of objects on Earth's surface and comparing it to satellite observations, scientists can accurately determine the rotational speed and axis of the Earth. These measurements align with the predictions of a round Earth and further debunk the flat Earth theory.

In conclusion, satellites and communication systems provide compelling evidence for the reality of our round Earth. The ability of satellites to function in specific orbits, the reliability of communication systems, and the data collected through satellite observations all support the scientific consensus on Earth's shape. The advancements in satellite technology have revolutionized our understanding of the world and continue to contribute to various fields of study.

7.4 Astronomical Observations and Space Missions

Astronomical observations and space missions have played a crucial role in providing evidence for the round Earth and debunking the flat Earth theory. Through these endeavors, scientists have been able to gather data and capture images that clearly demonstrate the true shape of our planet. In this section, we will explore some of the key astronomical observations and space missions that have contributed to our understanding of Earth's shape.

7.4.1 Lunar and Solar Eclipses

One of the most compelling pieces of evidence for a round Earth comes from the observation of lunar and solar eclipses. During a lunar eclipse, the Earth casts a shadow on the Moon, creating a reddish hue. The shape of this shadow is always round, regardless of the location from which the eclipse is observed. This phenomenon can only occur if the Earth is a sphere, as a flat Earth would cast a different shadow shape.

Similarly, during a solar eclipse, the Moon passes between the Earth and the Sun, causing a temporary blocking of sunlight. The shape of the shadow cast by the Moon on the Earth's surface is always round, regardless of the observer's location. This consistent round shadow is further evidence of the Earth's spherical shape.

7.4.2 Satellite Imagery

Satellites have provided us with a wealth of images that clearly show the Earth as a round object. Satellites such as NASA's Terra and Aqua, as well as international satellites like the European Space Agency's Sentinel-2, continuously capture images of the Earth from space. These images, freely available to the public, clearly depict a spherical Earth with a curvature that matches the predictions of a round planet.

Furthermore, satellite imagery has allowed us to observe the Earth from different angles and perspectives, providing a comprehensive view of our planet's shape. These images have become iconic symbols of our round Earth and have helped dispel the notion of a flat Earth.

7.4.3 Space Missions

Space missions have been instrumental in providing direct evidence of Earth's round shape. Astronauts who have traveled to space, such as those aboard the Apollo missions, have witnessed the curvature of the Earth firsthand. Their photographs and videos from space clearly show a spherical Earth, with a distinct horizon and a curvature that matches the predictions of a round planet.

In addition to manned missions, unmanned space probes have also contributed to our understanding of Earth's shape. For example, the Voyager 1 spacecraft, which was launched in 1977, captured the famous "Pale Blue Dot" image. This image, taken from a distance of about 6 billion kilometers away, shows Earth as a tiny, round speck in the vastness of space.

7.4.4 International Space Station (ISS)

The International Space Station (ISS) has been continuously occupied since November 2000 and has provided a unique platform for observing Earth from space. Astronauts aboard the ISS regularly capture images and videos of our planet, showcasing its round shape and the curvature of its surface.

The ISS orbits the Earth at an altitude of approximately 400 kilometers, allowing astronauts to witness breathtaking views of our planet. These images, shared with the public, serve as a powerful reminder of the reality of our round Earth.

7.4.5 GPS and Satellite Communication

The Global Positioning System (GPS) is a network of satellites that enables precise positioning and navigation on Earth. GPS relies on the concept of a round Earth to accurately calculate positions. The system works by triangulating signals from multiple satellites, which would not be possible if the Earth were flat.

Furthermore, satellite communication systems, such as those used for television broadcasts and internet connectivity, rely on the curvature of the Earth to transmit signals over long distances. These systems operate based on the understanding that the Earth is a round object, and their successful operation further supports the round Earth model.

In conclusion, astronomical observations and space missions have provided overwhelming evidence for the round Earth and have effectively debunked the flat Earth theory. Lunar and solar eclipses, satellite imagery, space missions, the International Space Station, GPS, and satellite communication systems all contribute to our understanding of Earth's shape. The wealth of data and images collected from these endeavors leave no doubt that our planet is indeed a sphere.

8 The Role of Science and Scientific Method

8.1 Understanding the Scientific Method

The scientific method is a systematic approach used by scientists to investigate and understand the natural world. It is a process that involves observation, experimentation, and analysis to develop and test hypotheses. By following this method, scientists can gather evidence and draw conclusions based on empirical data. In the context of the flat earth theory, understanding the scientific method is crucial in debunking the claims made by flat earthers and revealing the reality of our round earth.

8.1.1 Observation and Question

The scientific method begins with observation and questioning. Scientists carefully observe the natural world, identify patterns, and ask questions about how and why things happen. In the case of the flat earth theory, observations of the curvature of the earth, the round shadow cast during lunar eclipses, and the way objects disappear over the horizon raise questions about the validity of the flat earth model.

8.1.2 Formulating a Hypothesis

After making observations and asking questions, scientists formulate hypotheses. A hypothesis is a proposed explanation for a phenomenon based on existing knowledge and observations. In the case of the flat earth theory, the hypothesis would be that the earth is not flat but instead a spherical shape.

8.1.3 Testing the Hypothesis

Once a hypothesis is formulated, scientists design experiments and gather data to test its validity. In the case of the round earth hypothesis, numerous experiments and observations have been conducted to provide evidence supporting the spherical shape of the earth. These experiments include measuring the curvature of the earth, observing the behavior of gravity, and analyzing satellite imagery.

8.1.4 Analysis and Interpretation

After collecting data, scientists analyze and interpret the results to draw conclusions. This step involves statistical analysis, comparing the data to the initial hypothesis, and determining if the evidence supports or refutes the hypothesis. In the case of the round earth hypothesis, the analysis of various experiments and observations consistently supports the conclusion that the earth is indeed a sphere.

8.1.5 Peer Review and Scientific Consensus

One of the essential aspects of the scientific method is peer review. Scientists submit their findings to reputable scientific journals, where experts in the field review the research for accuracy, methodology, and validity. This rigorous process ensures that scientific claims are subjected to scrutiny and helps to establish scientific consensus. In the case of the round earth theory, the overwhelming consensus among scientists is that the earth is a sphere.

8.1.6 Reproducibility and Falsifiability

Reproducibility and falsifiability are crucial principles of the scientific method. For a scientific claim to be considered valid, it must be reproducible, meaning that other scientists should be able to replicate the experiments and obtain similar results. Additionally, scientific claims must be falsifiable, meaning that there must be a way to prove them wrong. In the case of the round earth theory, the experiments and observations supporting the spherical shape of the earth have been replicated by numerous scientists, further strengthening the validity of the claim.

8.1.7 The Flat Earth Theory and Scientific Scrutiny

The flat earth theory, despite its persistence, fails to meet the rigorous standards of the scientific method. The claims made by flat earthers lack empirical evidence, contradict well-established scientific principles, and are not supported by the scientific community. The scientific method provides a framework for testing and validating hypotheses, and the overwhelming evidence supports the conclusion that the earth is a sphere.

In conclusion, understanding the scientific method is essential in debunking the flat earth theory and revealing the reality of our round earth. By following the systematic approach of observation, hypothesis formulation, testing, analysis, and peer review, scientists have consistently provided evidence supporting the spherical shape of the earth. The scientific method ensures that claims are subjected to scrutiny, allowing for the advancement of knowledge and the rejection of unsupported theories.

8.2 Peer Review and Scientific Consensus

In the world of science, the process of peer review plays a crucial role in ensuring the quality and validity of research. Peer review involves the evaluation of scientific work by experts in the same field who assess the methodology, results, and conclusions of a study before it is published. This rigorous process helps to filter out flawed or biased research and ensures that only reliable and accurate information is disseminated to the scientific community and the public.

8.2.1 The Importance of Peer Review

Peer review serves as a critical checkpoint in the scientific method. It helps to maintain the integrity of scientific research by subjecting it to scrutiny from knowledgeable and impartial experts. When a study undergoes peer review, it is examined for its adherence to scientific principles, the validity of its data, the soundness of its methodology, and the logic of its conclusions.

The peer review process helps to identify any potential flaws or biases in a study, ensuring that any errors or inaccuracies are addressed before the research is published. It also provides an opportunity for experts in the field to offer constructive feedback and suggestions for improvement. This collaborative approach helps to refine scientific knowledge and advance the understanding of various phenomena.

8.2.2 Establishing Scientific Consensus

Scientific consensus refers to the collective agreement among experts in a particular field regarding a specific scientific theory or concept. It is reached through a rigorous evaluation of empirical evidence, repeated experimentation, and extensive peer review. Scientific consensus is not based on personal opinions or beliefs but on the weight of evidence and the consensus of experts.

The process of establishing scientific consensus involves the accumulation of a substantial body of evidence that consistently supports a particular theory or explanation. This evidence is then subjected to scrutiny by the scientific community through peer-reviewed publications, conferences, and discussions. Over time, as more evidence is gathered and analyzed, a consensus emerges among scientists who agree on the validity and reliability of a particular theory.

Scientific consensus is not static but can evolve as new evidence emerges or as existing theories are refined. However, it is important to note that scientific consensus is not easily swayed by individual opinions or unsubstantiated claims. It requires a robust and comprehensive body of evidence that can withstand scrutiny and replication by independent researchers.

8.2.3 The Flat Earth Theory and Peer Review

The flat earth theory, which posits that the Earth is a flat disc rather than a spherical planet, has been thoroughly examined and debunked by the scientific community. Numerous studies, experiments, and observations have provided overwhelming evidence in support of a round Earth. However, it is important to understand that the flat earth theory does not meet the rigorous standards of scientific peer review.

The claims made by flat earthers often lack empirical evidence, rely on flawed reasoning, and contradict well-established scientific principles. When subjected to peer review, these claims are consistently found to be unsupported by credible evidence and are deemed scientifically invalid. As a result, the flat earth theory remains outside the realm of scientific consensus.

It is worth noting that the scientific community is open to examining and evaluating any new evidence or theories that challenge existing knowledge. However, for a theory to gain acceptance, it must undergo the same rigorous process of peer review and scrutiny as any other scientific theory. The burden of proof lies with those proposing alternative explanations, and they must provide compelling evidence that withstands scientific scrutiny.

8.2.4 The Strength of Scientific Consensus

Scientific consensus is a powerful tool in distinguishing between valid scientific theories and pseudoscience. It represents the collective judgment of experts in a particular field and is based on the weight of evidence and the application of the scientific method. Scientific consensus provides a reliable foundation for understanding the natural world and informs public policy, technological advancements, and further scientific research.

While individual scientists may hold differing opinions or interpretations within a field, the consensus reflects the prevailing view supported by the majority of experts. It is important to recognize that scientific consensus is not infallible and can change as new evidence emerges. However, it is the most reliable and accurate representation of our current understanding of the natural world.

In the case of the shape of the Earth, the overwhelming scientific consensus supports the view that our planet is a sphere. This consensus is based on centuries of observations, geodetic surveys, satellite imagery, space exploration, and a multitude of other scientific evidence. The flat earth theory, lacking empirical evidence and scientific support, remains a fringe belief that stands in stark contrast to the scientific consensus.

In conclusion, peer review and scientific consensus are integral to the advancement of knowledge and the validation of scientific theories. The flat earth theory, lacking credible evidence and failing to meet the standards of scientific scrutiny, remains outside the realm of scientific consensus. The overwhelming body of evidence supports the reality of our round Earth, and it is through the rigorous application of the scientific method that we continue to uncover the truth about our planet and the universe.

8.3 Reproducibility and Falsifiability

In the realm of science, two fundamental principles play a crucial role in determining the validity and reliability of any scientific theory: reproducibility and falsifiability. These principles are essential for distinguishing between scientific claims that are supported by evidence and those that are mere speculation or pseudoscience. In this section, we will explore the concepts of reproducibility and falsifiability and their significance in the context of the flat earth theory.

8.3.1 Reproducibility: The Key to Scientific Validity

Reproducibility refers to the ability of an experiment or study to yield consistent results when repeated by different researchers or under different conditions. It is a cornerstone of the scientific method and serves as a critical test for the reliability and validity of scientific claims. In order for a scientific theory to be considered credible, its findings must be reproducible by independent researchers using the same methods and data.

When it comes to the flat earth theory, the lack of reproducibility is a significant issue. Flat earth proponents often rely on anecdotal evidence or personal experiences to support their claims, which makes it difficult for others to replicate their findings. Furthermore, the experiments and observations conducted by flat earth believers often lack rigorous methodology and fail to adhere to the standards of scientific inquiry.

Scientific experiments are designed to be reproducible, allowing other researchers to verify the results and build upon previous findings. However, in the case of the flat earth theory, the absence of reproducibility hinders the scientific community's ability to validate or refute the claims made by flat earth proponents. Without reproducible evidence, the flat earth theory remains unsupported by the scientific community.

8.3.2 Falsifiability: Challenging Scientific Theories

Falsifiability is another crucial principle in scientific inquiry. It refers to the ability of a scientific theory to be proven false through empirical evidence or experimentation. A falsifiable theory is one that makes specific predictions or claims that can be tested and potentially disproven. If a theory cannot be falsified, it falls outside the realm of science and enters the realm of pseudoscience or unfalsifiable beliefs.

The flat earth theory, unfortunately, lacks falsifiability. Flat earth proponents often rely on ad hoc explanations or unfalsifiable claims to counter any evidence that contradicts their beliefs. For example, when presented with photographs of the curved horizon or the spherical shape of the Earth from space, flat earth believers may dismiss them as part of a grand conspiracy or argue that the images have been doctored.

In contrast, the round earth theory is highly falsifiable. It makes specific predictions about the shape of the Earth, the curvature of the horizon, and the behavior of gravity, among other things. These predictions have been repeatedly tested and confirmed through various scientific experiments and observations. Any evidence that contradicts these predictions would potentially falsify the round earth theory and prompt a reevaluation of our understanding of the Earth's shape.

8.3.3 The Scientific Method and Flat Earth Claims

The scientific method is a systematic approach to acquiring knowledge and understanding the natural world. It involves formulating hypotheses, designing experiments, collecting data, analyzing results, and drawing conclusions. The scientific method relies on reproducibility and falsifiability to ensure the reliability and validity of scientific claims.

When examining the flat earth theory through the lens of the scientific method, it becomes clear that the theory falls short in meeting the criteria of reproducibility and falsifiability. The lack of rigorous experimentation, reliance on anecdotal evidence, and the dismissal of contradictory data undermine the scientific credibility of the flat earth theory.

Scientific theories, such as the round earth theory, have withstood rigorous scrutiny and have been repeatedly tested and confirmed through reproducible experiments and observations. The flat earth theory, on the other hand, fails to provide the necessary evidence and adherence to scientific principles to gain acceptance within the scientific community.

In conclusion, reproducibility and falsifiability are essential principles in scientific inquiry. The flat earth theory lacks reproducibility, as its claims are often based on anecdotal evidence and personal experiences that cannot be independently verified. Additionally, the theory lacks falsifiability, as it relies on unfalsifiable claims and ad hoc explanations to counter contradictory evidence. By contrast, the round earth theory has been extensively tested and confirmed through reproducible experiments and observations. The scientific method, with its emphasis on reproducibility and falsifiability, serves as a robust framework for evaluating scientific claims and distinguishing between credible scientific theories and pseudoscience.

8.4 The Flat Earth Theory and Scientific Scrutiny

The Flat Earth Theory has gained significant attention in recent years, fueled by the rise of social media and online communities. Despite overwhelming scientific evidence supporting the round Earth model, a small but vocal group of individuals continue to promote the idea that the Earth is flat. In this section, we will examine the claims made by flat Earthers and explore how scientific scrutiny debunks these assertions.

8.4.1 Misinterpretation of Observations

One of the main arguments put forth by flat Earthers is that the Earth appears flat to the naked eye. They claim that the horizon appears flat, and therefore, the Earth must be flat as well. However, this argument fails to consider the limitations of human perception and the vast scale of the Earth.

When standing on a flat surface, such as a large field or a calm body of water, the horizon does indeed appear flat. However, this is due to the curvature of the Earth being so subtle that it is not readily apparent to the naked eye. To truly observe the curvature, one needs to be at a much higher altitude or have access to satellite imagery.

8.4.2 Gravity and the Shape of the Earth

Another key aspect of the flat Earth theory is the rejection of gravity as a force that holds objects to the Earth's surface. Flat Earthers argue that objects should fall straight down if the Earth were truly round. However, this argument ignores the fundamental principles of gravity and the shape of the Earth.

Gravity is a force that pulls objects toward the center of mass. On a spherical Earth, this force acts perpendicular to the surface, causing objects to be pulled toward the center. This is why objects fall toward the ground and not in a straight line. The force of gravity is what gives the Earth its spherical shape.

8.4.3 Satellite Imagery and Space Exploration

Flat Earthers often dismiss satellite imagery and space exploration as elaborate hoaxes or fabrications. They claim that all images of the Earth from space are doctored or manipulated to fit the round Earth narrative. However, this argument is easily debunked by the wealth of evidence provided by space agencies and independent researchers.

Satellite imagery, such as those captured by NASA's Earth Observing System, clearly show the Earth as a round object. These images provide a comprehensive view of our planet, showing its curvature, weather patterns, and even the polar ice caps. Additionally, space missions, such as the Apollo moon landings, have provided astronauts with a firsthand perspective of the Earth's round shape.

8.4.4 Scientific Consensus and Peer Review

One of the cornerstones of the scientific method is the process of peer review. This rigorous evaluation ensures that scientific research is subjected to scrutiny by experts in the field before being accepted as valid. Flat Earthers often dismiss the scientific consensus on the round Earth as a result of a global conspiracy or bias.

However, the scientific consensus on the round Earth is not based on a single study or a handful of scientists. It is the result of centuries of research, observations, and experiments conducted by countless scientists from around the world. The evidence supporting the round Earth model is overwhelming and has been thoroughly reviewed and validated by experts in various scientific disciplines.

8.4.5 Debunking Flat Earth Claims

Scientists and researchers have dedicated significant time and effort to debunking the claims made by flat Earthers. Through careful analysis, experimentation, and observation, these claims have been consistently refuted.

From the curvature of the Earth to the behavior of gravity, the evidence overwhelmingly supports the round Earth model.

It is important to note that debunking flat Earth claims is not an attempt to ridicule or belittle those who hold these beliefs. Instead, it is a scientific endeavor to promote accurate understanding and knowledge about our planet. By addressing misconceptions and providing evidence-based explanations, we can encourage critical thinking and rationality in the face of misinformation.

In conclusion, the flat Earth theory does not withstand scientific scrutiny. The overwhelming evidence from geodetic surveys, satellite imagery, space exploration, and the scientific consensus all point to the Earth being a round object. By understanding the scientific method, embracing peer-reviewed research, and promoting scientific literacy, we can continue to uncover the reality of our round Earth and dispel the misconceptions perpetuated by the flat Earth theory.

9 Psychology and the Flat Earth Movement

9.1 Cognitive Biases and Belief Systems

Cognitive biases play a significant role in shaping our beliefs and perceptions of the world around us. They are inherent tendencies in human thinking that can lead us to make irrational judgments and hold onto false beliefs. When it comes to the flat earth movement, cognitive biases are often at the core of why individuals adhere to this belief system despite overwhelming evidence to the contrary.

9.1.1 Confirmation Bias

Confirmation bias is one of the most prevalent cognitive biases that influences our beliefs. It refers to our tendency to seek out and interpret information in a way that confirms our preexisting beliefs while ignoring or dismissing contradictory evidence. In the context of the flat earth theory, individuals who already hold this belief actively seek out information that supports their viewpoint, such as selective interpretation of scientific studies or anecdotal evidence. They may also dismiss or discredit any evidence that contradicts their belief, reinforcing their conviction in the flat earth theory.

9.1.2 Cognitive Dissonance

Cognitive dissonance occurs when there is a conflict between our beliefs and new information that challenges those beliefs. This discomfort can lead individuals to reject or rationalize the conflicting information in order to maintain consistency in their belief system. In the case of the flat earth theory, individuals may experience cognitive dissonance when confronted with scientific evidence that supports a round earth. To alleviate this discomfort, they may engage in mental gymnastics to reinterpret or dismiss the evidence, reinforcing their belief in a flat earth.

9.1.3 Dunning-Kruger Effect

The Dunning-Kruger effect refers to the tendency for individuals with low ability or knowledge in a particular area to overestimate their competence. In the context of the flat earth movement, some individuals may lack a deep understanding of scientific principles or the scientific method but still believe they possess superior knowledge on the subject. This overconfidence can lead them to dismiss the expertise of scientists and experts who have dedicated their lives to studying the shape of the earth. The Dunning-Kruger effect can create a false sense of certainty and reinforce their belief in the flat earth theory.

9.1.4 Backfire Effect

The backfire effect occurs when individuals, when presented with evidence that contradicts their beliefs, actually become more entrenched in those beliefs. This counterintuitive response is a result of the psychological discomfort caused by the challenge to their worldview. In the case of the flat earth theory, when presented with scientific evidence supporting a round earth, individuals may become defensive and double down on their belief in a flat earth. This can be attributed to the fear of admitting they were wrong or the desire to maintain their identity within the flat earth community.

9.1.5 Social Identity and Groupthink

Belonging to a community that shares a particular belief system can have a powerful influence on an individual's adherence to that belief. In the case of the flat earth movement, individuals often form strong social bonds within the community, which can create a sense of belonging and identity. This social identity can reinforce their belief in a flat earth and make it difficult for them to consider alternative viewpoints. Groupthink, a phenomenon where individuals prioritize conformity and consensus over critical thinking, can further solidify their beliefs and discourage dissenting opinions.

9.1.6 Availability Heuristic

The availability heuristic is a mental shortcut that relies on immediate examples that come to mind when evaluating a specific topic or making a judgment. In the case of the flat earth theory, individuals may rely on personal experiences or anecdotes that seem to support their belief, even if they are not representative of the larger body of evidence. For example, they may recall instances where the horizon appears flat or rely on personal observations that seem to contradict the round earth model. This reliance on easily accessible information can reinforce their belief in a flat earth, despite the overwhelming scientific evidence to the contrary.

9.1.7 Overcoming Cognitive Biases

Recognizing and understanding cognitive biases is crucial for critical thinking and rational decision-making. Overcoming these biases requires a willingness to challenge our own beliefs and actively seek out diverse perspectives and evidence. Engaging in open-minded discussions, evaluating the credibility of sources, and being receptive to new information are essential steps in overcoming cognitive biases. Additionally, promoting scientific literacy and education can help individuals develop the necessary skills to critically evaluate evidence and make informed judgments.

In conclusion, cognitive biases play a significant role in shaping and reinforcing beliefs within the flat earth movement. Confirmation bias, cognitive dissonance, the Dunning-Kruger effect, the backfire effect, social identity, groupthink, the availability heuristic, and other biases contribute to the persistence of the flat earth theory despite overwhelming scientific evidence supporting a round earth. Understanding these biases is crucial in promoting critical thinking, scientific literacy, and rationality in our society.

9.2 Conspiracy Theories and Confirmation Bias

Conspiracy theories have always been a part of human history, and the flat earth theory is no exception. The belief in a flat earth is often intertwined with various conspiracy theories, which serve to reinforce and perpetuate the idea. One of the psychological factors that contribute to the persistence of these beliefs is confirmation bias.

Confirmation bias is the tendency to seek out and interpret information in a way that confirms one's preexisting beliefs or hypotheses. In the case of flat earth believers, confirmation bias plays a significant role in their acceptance and propagation of conspiracy theories. When confronted with evidence that contradicts their beliefs, they often dismiss it or reinterpret it to fit their worldview.

9.2.1 The Conspiracy of Suppression

One of the most common conspiracy theories among flat earth proponents is the idea that there is a vast global conspiracy to suppress the truth about the flat earth. According to this theory, governments, space agencies, and scientific institutions are all involved in a grand cover-up to maintain the illusion of a round earth.

Believers in this conspiracy argue that any evidence or scientific research that supports the flat earth theory is intentionally suppressed or discredited. They claim that scientists who speak out in favor of a flat earth are silenced or ostracized, and that any photographs or videos showing a flat horizon are doctored or manipulated.

9.2.2 Confirmation Bias in Action

Confirmation bias plays a crucial role in perpetuating the conspiracy of suppression. Flat earth believers actively seek out information that supports their preexisting beliefs while disregarding or dismissing any evidence to the contrary. They often rely on anecdotal accounts, personal experiences, and selective interpretation of data to confirm their worldview.

For example, when presented with satellite imagery showing a round earth, flat earth proponents may claim that the images are part of the conspiracy and that the curvature of the earth is merely an optical illusion. They may also dismiss scientific experiments and measurements as flawed or biased, reinforcing their belief that the truth is being hidden from the public.

9.2.3 The Echo Chamber Effect

Another factor that contributes to the persistence of flat earth beliefs is the echo chamber effect. In today's digital age, like-minded individuals can easily find online communities and social media groups that reinforce their beliefs. Within these echo chambers, confirmation bias is amplified as members share and validate each other's views, creating an environment where dissenting opinions are often dismissed or ridiculed.

The echo chamber effect can be particularly powerful in shaping individuals' beliefs and attitudes. When surrounded by a community that shares the same beliefs, it becomes increasingly difficult for individuals to question or critically evaluate their own ideas. This reinforces confirmation bias and further entrenches the belief in a flat earth.

9.2.4 Overcoming Confirmation Bias

Overcoming confirmation bias is challenging, but not impossible. It requires individuals to be open-minded, willing to critically evaluate evidence, and consider alternative viewpoints. It is essential to approach information with skepticism and seek out reliable sources that are based on scientific evidence and consensus.

Engaging in respectful and constructive discussions with individuals who hold different beliefs can also help challenge confirmation bias. By listening to opposing viewpoints and examining the evidence presented, individuals can gain a more comprehensive understanding of the topic and potentially reevaluate their own beliefs.

Furthermore, promoting scientific literacy and critical thinking skills is crucial in combating confirmation bias. By educating individuals about the scientific method, logical reasoning, and the importance of evidence-based thinking, we can empower them to make informed decisions and resist the influence of confirmation bias.

In conclusion, conspiracy theories and confirmation bias play a significant role in perpetuating the belief in a flat earth. The conspiracy of suppression and the echo chamber effect contribute to the persistence of these beliefs, making it challenging for individuals to consider alternative viewpoints or evaluate evidence objectively. Overcoming confirmation bias requires open-mindedness, critical thinking, and a commitment to scientific literacy. By addressing these psychological factors, we can encourage a more rational and evidence-based understanding of our round earth.

9.3 Social Dynamics and Groupthink

In the previous sections, we have explored the cognitive biases and belief systems that contribute to the persistence of the flat earth theory. However, it is important to also examine the social dynamics and groupthink that play a significant role in the spread and perpetuation of this belief system.

9.3.1 The Power of Social Influence

Humans are social beings, and our beliefs and behaviors are often influenced by the people around us. This is particularly true when it comes to controversial or fringe beliefs like the flat earth theory. Social influence can take various forms, including direct persuasion, conformity, and the desire for social acceptance.

One of the key factors contributing to the spread of the flat earth theory is the existence of online communities and social media platforms that provide a sense of belonging and validation for flat earthers. These communities create an echo chamber where like-minded individuals reinforce each other's beliefs and dismiss any contradictory evidence or arguments.

9.3.2 Group Polarization and Confirmation Bias

Group polarization is a phenomenon where the attitudes and beliefs of a group become more extreme over time. In the context of the flat earth movement, this means that when individuals with similar beliefs come together, their collective conviction in the flat earth theory intensifies. This polarization is fueled by confirmation bias, which is the tendency to seek out and interpret information in a way that confirms preexisting beliefs.

Within flat earth communities, group polarization and confirmation bias create a reinforcing cycle. As members share and discuss their beliefs, they become more convinced of the validity of the flat earth theory. Any evidence or arguments that contradict their beliefs are dismissed or rationalized away, further strengthening their conviction.

9.3.3 Identity and Belonging

Belonging to a group that shares a fringe belief like the flat earth theory can provide individuals with a sense of identity and purpose. It offers a community where they feel understood and accepted, which can be particularly appealing for those who feel marginalized or alienated from mainstream society. The flat earth movement becomes not just a belief system but also a part of their personal identity.

This sense of identity and belonging can make it incredibly difficult for individuals to question or reconsider their beliefs. Doing so would mean risking the loss of their social connections and the support system they have found within the flat earth community. As a result, many flat earthers become deeply entrenched in their beliefs, resisting any attempts to challenge or debunk them.

9.3.4 Authority Figures and Charismatic Leaders

The flat earth movement, like many other conspiracy theories, often has charismatic leaders or influential figures who play a crucial role in shaping and spreading the belief system. These individuals may have a strong presence on social media platforms or in online communities, and their followers often view them as experts or authorities on the subject.

The influence of these authority figures is significant, as their followers tend to trust and rely on their guidance. They provide a sense of credibility and validation for the flat earth theory, reinforcing the beliefs of their followers.

This dynamic can make it even more challenging for individuals within the movement to question or critically evaluate the evidence presented to them.

9.3.5 Overcoming Groupthink

Overcoming the social dynamics and groupthink associated with the flat earth movement is a complex task. It requires a multifaceted approach that addresses both the individual and collective aspects of belief formation and maintenance.

One potential strategy is to foster critical thinking skills and scientific literacy. By promoting a deeper understanding of the scientific method and encouraging individuals to question and evaluate evidence independently, we can empower them to think critically about the flat earth theory and other fringe beliefs.

Additionally, creating spaces for open dialogue and respectful debate can help challenge the echo chamber effect within flat earth communities. By exposing individuals to diverse perspectives and evidence, we can encourage them to consider alternative viewpoints and potentially reevaluate their beliefs.

It is also important to recognize the emotional and psychological factors that contribute to the adoption of fringe beliefs. Providing support and empathy to individuals who may be struggling with their beliefs can help create an environment where questioning and growth are welcomed.

Ultimately, addressing the social dynamics and groupthink associated with the flat earth movement requires a combination of education, empathy, and open dialogue. By understanding the underlying factors that contribute to the persistence of this belief system, we can work towards promoting a more rational and evidence-based understanding of our round earth.

9.4 Debunking Flat Earth Psychological Factors

While scientific evidence overwhelmingly supports the round Earth model, the persistence of the flat Earth theory is intriguing. It is important to understand the psychological factors that contribute to the belief in a flat Earth and to debunk them with rational and evidence-based arguments. In this section, we will explore some of the key psychological factors that contribute to the flat Earth movement and provide counterarguments to debunk them.

9.4.1 Cognitive Dissonance

Cognitive dissonance refers to the discomfort experienced when holding conflicting beliefs or when faced with information that contradicts one's existing beliefs. Flat Earth proponents often experience cognitive dissonance when confronted with scientific evidence that supports a round Earth. To alleviate this discomfort, they may reject or dismiss the evidence, clinging to their preconceived notions.

To debunk this psychological factor, it is crucial to present the scientific evidence in a clear and accessible manner. By providing compelling explanations and empirical data, we can help individuals reconcile their conflicting beliefs and overcome cognitive dissonance.

9.4.2 Confirmation Bias

Confirmation bias is the tendency to seek out and interpret information in a way that confirms one's preexisting beliefs while disregarding contradictory evidence. Flat Earth believers often engage in selective information processing, actively seeking out sources that support their views and dismissing any evidence that contradicts their beliefs.

To counter confirmation bias, it is essential to encourage critical thinking and promote the evaluation of evidence from multiple perspectives. By presenting a comprehensive range of scientific studies and observations, we can challenge individuals to question their biases and consider alternative explanations.

9.4.3 Distrust in Authority

Many flat Earth proponents exhibit a deep-seated distrust in authority figures, including scientists, government institutions, and educational systems. This distrust often stems from a belief in widespread conspiracies and a perception that these authorities are intentionally hiding the truth about the shape of the Earth.

To address this psychological factor, it is important to emphasize the rigorous nature of scientific inquiry and the transparency of the scientific process. By highlighting the peer-review system, the reproducibility of experiments, and the consensus-building nature of scientific knowledge, we can help individuals understand that scientific findings are not the result of a grand conspiracy but rather the product of rigorous investigation.

9.4.4 Social Identity and Group Dynamics

Belief in a flat Earth often becomes intertwined with one's social identity, leading to a sense of belonging within the flat Earth community. This sense of belonging can create a strong emotional attachment to the belief system, making it difficult for individuals to consider alternative viewpoints.

To counteract the influence of social identity and group dynamics, it is important to foster an environment that encourages open-mindedness and critical thinking. By promoting respectful dialogue and providing opportunities for individuals to engage with diverse perspectives, we can help individuals break free from the echo chambers of the flat Earth community and consider the overwhelming evidence for a round Earth.

9.4.5 Lack of Scientific Literacy

A lack of scientific literacy is a significant contributing factor to the acceptance of flat Earth beliefs. Many individuals may not have a solid understanding of scientific principles, making them more susceptible to misinformation and pseudoscience.

To address this issue, it is crucial to prioritize science education and promote scientific literacy. By providing accessible and engaging educational resources, we can empower individuals to evaluate scientific claims critically and make informed decisions based on evidence.

9.4.6 Emotional and Personal Factors

Emotional and personal factors, such as fear, mistrust, and the desire for uniqueness, can also contribute to the acceptance of flat Earth beliefs. These factors may be deeply rooted in an individual's personal experiences or psychological needs.

To debunk these emotional and personal factors, it is important to approach discussions with empathy and understanding. By addressing the underlying emotional needs and providing reassurance that accepting the round Earth model does not diminish one's uniqueness or personal experiences, we can help individuals overcome these barriers and embrace scientific reality.

In conclusion, debunking the psychological factors that contribute to the flat Earth movement requires a multifaceted approach. By addressing cognitive dissonance, confirmation bias, distrust in authority, social identity, lack of scientific literacy, and emotional/personal factors, we can help individuals overcome their beliefs in a flat Earth and embrace the overwhelming evidence for a round Earth. Through education, critical thinking, and open dialogue, we can promote a society that embraces scientific literacy and rationality.

10 Educating and Communicating Science

10.1 Challenges in Communicating Complex Science

Communicating complex scientific concepts can be a challenging task, especially when it comes to debunking misconceptions such as the flat earth theory. The flat earth movement has gained traction in recent years, fueled by the spread of misinformation and the ease of access to alternative viewpoints on the internet. As a result, scientists and educators face numerous challenges in effectively communicating the reality of our round earth. In this section, we will explore some of these challenges and discuss strategies for overcoming them.

10.1.1 Lack of Scientific Literacy

One of the primary challenges in communicating complex science is the widespread lack of scientific literacy. Many individuals may not have a solid understanding of basic scientific principles, making it difficult for them to grasp more complex concepts such as the shape of the earth. This lack of scientific literacy can be attributed to various factors, including inadequate science education, misinformation, and the prevalence of pseudoscience.

To address this challenge, it is crucial to start by building a strong foundation of scientific knowledge. Educators and science communicators should focus on promoting scientific literacy from an early age, emphasizing critical thinking, evidence-based reasoning, and the scientific method. By fostering a solid understanding of scientific principles, individuals are better equipped to evaluate and comprehend complex scientific concepts.

10.1.2 Cognitive Biases and Confirmation Bias

Another challenge in communicating complex science is the presence of cognitive biases, particularly confirmation bias. Confirmation bias refers to the tendency of individuals to seek out and interpret information in a way that confirms their preexisting beliefs or biases. In the case of the flat earth theory, individuals who subscribe to this belief may actively seek out information that supports their viewpoint while dismissing or ignoring contradictory evidence.

To overcome confirmation bias, it is essential to approach science communication with empathy and understanding. Instead of directly challenging individuals' beliefs, it is more effective to present them with compelling evidence and encourage critical thinking. By creating a safe and non-confrontational environment, individuals may be more open to considering alternative viewpoints and reevaluating their beliefs.

10.1.3 Overcoming Skepticism and Distrust

Communicating complex science also requires addressing skepticism and distrust. The flat earth movement often thrives on conspiracy theories and the notion that scientific institutions are engaged in a grand cover-up. This skepticism can make it challenging to establish trust and credibility when presenting scientific evidence.

To overcome this challenge, it is crucial to emphasize the transparency and openness of the scientific process. Scientists and educators should actively engage with the public, sharing their research methodologies, data, and findings. By promoting transparency and inviting public scrutiny, individuals may develop a greater sense of trust in the scientific community.

10.1.4 Tailoring Communication to Different Audiences

Effective science communication requires tailoring the message to different audiences. Not everyone has the same level of scientific background or understanding, and using technical jargon or complex terminology can alienate individuals and hinder comprehension.

To address this challenge, it is important to adapt the communication style to suit the target audience. Using clear and concise language, avoiding jargon, and utilizing relatable examples can help make complex scientific concepts more accessible. Additionally, employing visual aids, such as diagrams or illustrations, can enhance understanding and engagement.

10.1.5 Addressing Emotional and Personal Beliefs

Communicating complex science often involves addressing deeply held emotional and personal beliefs. The flat earth theory, for example, is often intertwined with individuals' identities and worldviews. Challenging these beliefs can evoke strong emotional responses, making it difficult to have productive conversations.

To navigate this challenge, it is important to approach discussions with empathy and respect. Acknowledging individuals' emotions and validating their experiences can help create a more conducive environment for dialogue. It is also crucial to focus on the evidence and scientific consensus rather than engaging in personal attacks or dismissive behavior.

10.1.6 Leveraging Multiple Communication Channels

In the age of information, utilizing multiple communication channels is essential for reaching a wider audience. Traditional methods such as books, articles, and lectures are still valuable, but digital platforms offer unique opportunities for engaging with diverse audiences.

Science communicators should leverage social media, podcasts, videos, and interactive websites to disseminate scientific information effectively. These platforms allow for greater accessibility, interactivity, and engagement. By utilizing a variety of communication channels, scientists and educators can reach individuals who may not have access to traditional educational resources.

In conclusion, communicating complex science, particularly when debunking misconceptions like the flat earth theory, presents numerous challenges. Overcoming these challenges requires promoting scientific literacy, addressing cognitive biases and skepticism, tailoring communication to different audiences, addressing emotional beliefs, and leveraging multiple communication channels. By employing effective science communication strategies, we can foster a better understanding of the reality of our round earth and promote scientific literacy in society.

10.2 Effective Science Communication Strategies

In order to effectively communicate the reality of our round Earth and debunk the flat Earth theory, it is crucial to employ effective science communication strategies. The flat Earth theory has gained traction in recent years, fueled by misinformation and conspiracy theories. To counteract this, scientists and educators must employ strategies that are engaging, accessible, and based on scientific evidence. Here are some effective science communication strategies to address misconceptions and skepticism surrounding the shape of our planet.

10.2.1 Simplify Complex Concepts

One of the challenges in communicating complex science is the use of technical jargon and complicated explanations. To effectively communicate the reality of our round Earth, it is important to simplify complex concepts and present them in a way that is easily understandable to a wide audience. Using analogies, visual aids, and relatable examples can help make scientific concepts more accessible and relatable.

For example, when explaining the curvature of the Earth, instead of diving into mathematical formulas, one can use the analogy of a basketball. By comparing the Earth to a basketball, it becomes easier for people to visualize the curvature and understand that it is not flat.

10.2.2 Tailor the Message to the Audience

Different audiences have different levels of scientific literacy and varying degrees of familiarity with the flat Earth theory. It is important to tailor the message to the specific audience in order to effectively communicate the scientific evidence. For example, when addressing a general audience, it may be necessary to start with basic concepts and gradually build up to more complex ideas. On the other hand, when engaging with individuals who are already familiar with the flat Earth theory, it may be more effective to directly address their specific misconceptions and provide evidence to counter them.

10.2.3 Use Engaging and Interactive Methods

Engaging and interactive methods of science communication can be highly effective in capturing the attention and interest of the audience. Utilizing multimedia platforms such as videos, animations, and interactive websites can help convey scientific information in a visually appealing and interactive manner. These methods can make the learning experience more enjoyable and memorable, increasing the likelihood that the audience will retain and understand the information being presented.

For example, creating a virtual reality experience that allows individuals to explore the Earth from space can provide a unique perspective and help them understand the spherical nature of our planet.

10.2.4 Incorporate Personal Stories and Testimonials

In addition to presenting scientific evidence, incorporating personal stories and testimonials can help humanize the message and make it more relatable. Sharing stories of individuals who were once convinced by the flat Earth theory but later changed their beliefs after examining the scientific evidence can be powerful in challenging misconceptions and inspiring critical thinking.

10.2.5 Foster Open Dialogue and Address Concerns

When communicating the reality of our round Earth, it is important to foster open dialogue and address concerns and questions from the audience. Encouraging individuals to ask questions and providing clear and concise answers can help build trust and credibility. It is essential to approach these conversations with patience, empathy, and respect, as individuals may have deeply ingrained beliefs and may be resistant to change.

10.2.6 Collaborate with Influencers and Opinion Leaders

In the age of social media, collaborating with influencers and opinion leaders who have a large following can significantly amplify the reach and impact of science communication efforts. By partnering with individuals who have credibility and influence within the target audience, scientists and educators can effectively disseminate accurate information and counteract the spread of misinformation.

10.2.7 Utilize Social Media and Online Platforms

Social media and online platforms provide powerful tools for science communication. Utilizing platforms such as YouTube, Twitter, and Facebook can help reach a wide audience and engage with individuals who may be exposed to flat Earth theories. Creating informative and visually appealing content, participating in online discussions, and sharing scientific evidence can help counteract misinformation and promote scientific literacy.

10.2.8 Address Emotional and Psychological Factors

When communicating the reality of our round Earth, it is important to acknowledge and address the emotional and psychological factors that may contribute to the belief in the flat Earth theory. Many individuals who subscribe to this theory may have deeply held beliefs and may feel a sense of belonging within the flat Earth community. It is crucial to approach these conversations with empathy and understanding, while still presenting the scientific evidence in a clear and concise manner.

By employing these effective science communication strategies, scientists and educators can effectively debunk the flat Earth theory and promote scientific literacy. It is important to remember that changing deeply held beliefs takes time and patience. By engaging in open dialogue, presenting evidence-based arguments, and fostering critical thinking, we can help individuals embrace the reality of our round Earth and navigate the age of information with a strong foundation in scientific understanding.

10.3 Addressing Misconceptions and Skepticism

In the previous chapters, we have explored the scientific evidence that supports the reality of our round Earth. However, despite the overwhelming evidence, there are still individuals who hold onto the belief in a flat Earth. Addressing misconceptions and skepticism is crucial in promoting scientific literacy and dispelling misinformation. In this section, we will delve into some of the common misconceptions and skepticism surrounding the round Earth theory and provide evidence to counter these claims.

10.3.1 Misconception: "The Horizon Always Appears Flat"

One of the most common misconceptions is that the horizon always appears flat, therefore suggesting a flat Earth. However, this misconception arises from a limited perspective and a lack of understanding of how our eyes perceive the curvature of the Earth. From ground level, the curvature is not readily apparent due to the vast size of our planet. However, as we gain altitude, such as from an airplane or a high vantage point, the curvature becomes more visible.

Furthermore, the curvature of the Earth can be observed in various ways. For example, when watching a ship sail away from the shore, it gradually disappears from view, with the hull being the last part to vanish. This phenomenon, known as "the sinking ship effect," is a direct result of the Earth's curvature. Similarly, when observing the sunset or sunrise, the bottom part of the Sun appears to rise or set first, while the top part remains visible for a longer period. These observations provide clear evidence of the Earth's curvature.

10.3.2 Misconception: "Gravity Cannot Explain Why We Don't Fall Off the Earth"

Another misconception often raised by flat Earth proponents is the idea that gravity cannot explain why we don't fall off the Earth. They argue that if the Earth were truly round, gravity would cause us to fall off the surface. However, this misconception stems from a misunderstanding of how gravity works.

Gravity is the force that attracts objects towards the center of mass. On a round Earth, gravity pulls everything towards the center, creating a downward force that keeps us firmly on the ground. The force of gravity is proportional to the mass of an object, meaning that larger objects, like the Earth, have a stronger gravitational pull. This is why we don't float away into space.

Additionally, the force of gravity decreases with distance from the center of mass. As we move away from the Earth's surface, the force of gravity weakens, allowing objects like satellites to orbit the Earth. This is why astronauts in the International Space Station experience a microgravity environment. The concept of gravity is well-supported by scientific evidence and has been extensively studied and understood for centuries.

10.3.3 Skepticism: "The Conspiracy Theory of a Round Earth"

Some skeptics of the round Earth theory propose that the belief in a spherical planet is a result of a vast conspiracy involving governments, scientists, and various organizations. They argue that these entities are intentionally hiding the truth about the Earth's shape for unknown reasons.

However, the idea of a global conspiracy involving millions of people across different countries and organizations is highly implausible. It would require an unprecedented level of coordination and secrecy, which is simply not feasible. Moreover, the evidence supporting the round Earth theory comes from a wide range of sources, including independent scientific research, space exploration, and satellite imagery. These pieces of evidence have been verified and scrutinized by countless scientists and experts worldwide.

Scientific knowledge is built upon the principles of peer review, reproducibility, and falsifiability. Any evidence or theory that contradicts the round Earth model would be subject to rigorous scrutiny and testing. The scientific community thrives on skepticism and critical thinking, and any evidence supporting a flat Earth would be welcomed and thoroughly examined. However, to date, no credible scientific evidence has been presented to challenge the overwhelming consensus that the Earth is round.

10.3.4 Skepticism: "The Lack of Personal Observation"

Another common skepticism raised by flat Earth proponents is the idea that individuals cannot personally observe the curvature of the Earth in their day-to-day lives. They argue that since most people do not have access to high-altitude vantage points or space travel, they cannot verify the round Earth for themselves.

While it is true that most people do not have the opportunity to observe the curvature firsthand, this does not invalidate the scientific evidence that supports the round Earth theory. Science is not solely based on personal observation but relies on a collective body of knowledge and evidence gathered through rigorous experimentation and observation by experts in the field.

Furthermore, the availability of satellite imagery, photographs taken from space, and the testimonies of astronauts who have seen the Earth from orbit provide ample evidence of our planet's round shape. These sources of evidence are accessible to anyone with an internet connection, allowing individuals to explore and understand the reality of our round Earth.

In conclusion, addressing misconceptions and skepticism is essential in promoting scientific literacy and dispelling misinformation. By providing evidence and explanations, we can counter the claims of the flat Earth theory and encourage critical thinking and rationality. The overwhelming scientific consensus, supported by centuries of research and observation, confirms that our Earth is indeed round.

10.4 Promoting Scientific Literacy

Scientific literacy is crucial in today's world, where misinformation and pseudoscience can easily spread and influence public opinion. It is essential to promote scientific literacy to combat the spread of flat earth beliefs and other unfounded theories. By equipping individuals with the necessary knowledge and critical thinking skills, we can empower them to make informed decisions based on evidence and reason. In this section, we will explore the importance of promoting scientific literacy and discuss strategies to achieve this goal.

10.4.1 The Importance of Scientific Literacy

Scientific literacy refers to the ability to understand and evaluate scientific concepts, theories, and evidence. It involves having a basic understanding of the scientific method, critical thinking skills, and the ability to distinguish between reliable scientific information and pseudoscience. Scientific literacy is not limited to scientists; it is a skill that everyone should possess to navigate the complex world we live in.

Promoting scientific literacy is essential for several reasons. Firstly, it enables individuals to make informed decisions about their health, the environment, and other important aspects of their lives. Scientific literacy empowers individuals to critically evaluate claims and make choices based on evidence rather than misinformation or personal beliefs.

Secondly, scientific literacy fosters a deeper appreciation and understanding of the natural world. It allows individuals to marvel at the wonders of the universe, from the intricate workings of the human body to the vastness of space. By promoting scientific literacy, we can inspire curiosity and a lifelong love for learning.

Lastly, scientific literacy is crucial for the progress of society. It drives innovation, technological advancements, and economic growth. A scientifically literate population can contribute to solving complex problems, such as climate change, by understanding the underlying scientific principles and supporting evidence-based policies.

10.4.2 Strategies for Promoting Scientific Literacy

Promoting scientific literacy requires a multi-faceted approach that involves educators, scientists, policymakers, and the media. Here are some strategies that can be employed to enhance scientific literacy:

10.4.2.1 Science Education Reform

Science education plays a fundamental role in promoting scientific literacy. It is essential to reform science curricula to emphasize critical thinking, hands-on experimentation, and the understanding of scientific principles. By incorporating real-world examples and engaging students in scientific inquiry, educators can foster a deeper understanding and appreciation for science.

10.4.2.2 Public Engagement and Outreach

Scientists and science communicators have a responsibility to engage with the public and communicate scientific concepts effectively. Public lectures, science festivals, and interactive exhibits can help bridge the gap between scientists and the general public. By making science accessible and relatable, we can inspire curiosity and promote scientific literacy.

10.4.2.3 Media Literacy

In the age of information, media literacy is crucial for distinguishing between reliable scientific information and misinformation. Educating individuals on how to critically evaluate sources, fact-check claims, and recognize bias can empower them to make informed decisions. Media organizations should also prioritize accurate and evidence-based reporting to promote scientific literacy.

10.4.2.4 Collaboration between Scientists and Educators

Collaboration between scientists and educators can enhance science education and promote scientific literacy. Scientists can provide expertise, resources, and real-world examples to enrich classroom learning. By sharing their research and experiences, scientists can inspire students and foster a deeper understanding of scientific concepts.

10.4.2.5 Promoting Critical Thinking

Critical thinking is a fundamental skill for scientific literacy. Educators should incorporate activities that encourage students to question, analyze evidence, and evaluate arguments. By teaching students how to think critically, we can equip them with the tools to navigate the vast amount of information available and make informed decisions.

10.4.3 Overcoming Challenges

Promoting scientific literacy faces several challenges. Misinformation, confirmation bias, and the spread of pseudoscience can hinder efforts to educate the public. Overcoming these challenges requires a concerted effort from scientists, educators, policymakers, and the media.

To combat misinformation, it is crucial to provide accessible and accurate scientific information through various channels. Scientists and science communicators should actively engage with the public, addressing misconceptions and debunking unfounded claims. Fact-checking organizations and reputable scientific institutions can play a vital role in providing reliable information to counteract misinformation.

Addressing confirmation bias requires fostering an environment that encourages open-mindedness and critical thinking. Educators should create spaces where students feel comfortable questioning and challenging ideas. Encouraging dialogue and respectful debate can help individuals overcome their biases and develop a more nuanced understanding of scientific concepts.

The media also plays a significant role in promoting scientific literacy. Journalists should prioritize accurate reporting and avoid sensationalism. Media organizations can collaborate with scientists to ensure that scientific information is communicated effectively and accurately to the public.

10.4.4 The Role of Individuals

Promoting scientific literacy is not solely the responsibility of educators, scientists, and policymakers. Individuals also have a role to play in enhancing their own scientific literacy and that of their communities. Here are some steps individuals can take:

- Seek out reliable sources of scientific information and fact-check claims before accepting them.
- Engage in critical thinking and question the validity of information.
- Stay informed about scientific advancements and discoveries through reputable sources.
- Support science education initiatives and advocate for evidence-based policies.
- Encourage open dialogue and respectful discussions about scientific topics with friends and family.

By taking these steps, individuals can contribute to the promotion of scientific literacy and help combat the spread of unfounded beliefs such as the flat earth theory.

In conclusion, promoting scientific literacy is crucial in today's world. By equipping individuals with the necessary knowledge and critical thinking skills, we can combat the spread of misinformation and pseudoscience. Through science education reform, public engagement, media literacy, collaboration, and promoting critical thinking, we can enhance scientific literacy and empower individuals to make informed decisions based on evidence and reason. It is a collective effort that requires the involvement of educators, scientists, policymakers, the media, and individuals themselves.

Together, we can build a society that embraces the reality of our round earth and values scientific literacy.

11 The Impact of Flat Earth Beliefs

11.1 Societal and Educational Consequences

The belief in a flat Earth has far-reaching consequences for society and education. While it may seem harmless to some, the promotion of flat Earth beliefs can have detrimental effects on individuals and the broader community. In this section, we will explore the societal and educational consequences of embracing flat Earth beliefs.

11.1.1 Undermining Scientific Understanding

One of the most significant consequences of flat Earth beliefs is the erosion of scientific understanding. Science is a systematic and evidence-based approach to understanding the natural world. It relies on observation, experimentation, and peer review to develop accurate explanations of phenomena. By rejecting the overwhelming scientific evidence supporting a round Earth, flat Earth beliefs undermine the very foundation of scientific knowledge.

When individuals embrace flat Earth beliefs, they often reject established scientific principles and methodologies. This rejection can lead to a distrust of scientific institutions and experts, hindering progress in various fields of study. It becomes challenging to engage in meaningful scientific discourse when a fundamental aspect of our understanding of the Earth is dismissed.

11.1.2 Misinformation and Pseudoscience

The spread of flat Earth beliefs contributes to the proliferation of misinformation and pseudoscience. Misinformation refers to false or misleading information that is spread unintentionally, while pseudoscience refers to claims that are presented as scientific but lack empirical evidence or fail to adhere to scientific principles.

Flat Earth beliefs often rely on cherry-picked data, logical fallacies, and conspiracy theories to support their claims. This misinformation can be easily spread through social media platforms and other online channels, leading to the creation of echo chambers where individuals reinforce each other's misconceptions. The acceptance of pseudoscience and misinformation can have severe consequences, as it hampers critical thinking and the ability to discern reliable information from falsehoods.

11.1.3 Educational Challenges

The prevalence of flat Earth beliefs poses significant challenges for educators. In an era where scientific literacy is crucial, the promotion of flat Earth beliefs can hinder the educational process. Teachers must navigate the delicate balance between respecting students' beliefs and providing accurate scientific information.

When students are exposed to flat Earth beliefs, it can lead to confusion and a lack of trust in educational institutions. It becomes challenging for educators to teach scientific concepts such as gravity, Earth's shape, and the solar system when students are exposed to alternative explanations that lack scientific validity. This can hinder the development of critical thinking skills and the ability to evaluate evidence objectively.

11.1.4 Hindering Technological Advancements

The acceptance of flat Earth beliefs can hinder technological advancements and scientific progress. Many technological innovations, such as GPS systems, satellite communication, and space exploration, rely on our understanding of a round Earth. By rejecting this understanding, individuals limit their ability to fully engage with and benefit from these advancements.

Furthermore, the promotion of flat Earth beliefs can discourage young minds from pursuing careers in science, technology, engineering, and mathematics

(STEM) fields. The rejection of scientific principles and the dismissal of evidence-based reasoning can discourage individuals from pursuing scientific inquiry and contributing to technological advancements.

11.1.5 Polarizing Society

Flat Earth beliefs can also contribute to the polarization of society. The promotion of alternative explanations for well-established scientific facts can create divisions between those who accept scientific consensus and those who reject it. This polarization can lead to a breakdown in communication, increased distrust, and the formation of echo chambers where individuals reinforce their own beliefs without engaging with opposing viewpoints.

The societal consequences of flat Earth beliefs extend beyond the realm of science. They can impact public policy decisions, environmental awareness, and even social cohesion. It is essential to recognize the potential harm that can arise from the spread of misinformation and pseudoscience and work towards promoting a scientifically literate society.

In conclusion, the societal and educational consequences of flat Earth beliefs are significant. They undermine scientific understanding, contribute to the spread of misinformation and pseudoscience, pose challenges for educators, hinder technological advancements, and polarize society. It is crucial to address these consequences by promoting scientific literacy, critical thinking, and rationality. By embracing the reality of our round Earth, we can foster a society that values evidence-based reasoning and engages in meaningful scientific discourse.

11.2 Pseudoscience and Misinformation

In the age of information, where knowledge is readily accessible at our fingertips, it is disheartening to see the rise of pseudoscience and misinformation. The flat earth theory is a prime example of how misinformation can spread and gain traction, despite overwhelming evidence to the contrary. This chapter aims to shed light on the dangers of pseudoscience and misinformation, particularly in relation to the flat earth movement.

11.2.1 The Spread of Pseudoscience

Pseudoscience refers to beliefs or practices that are presented as scientific but lack empirical evidence, logical reasoning, or adherence to the scientific method. The flat earth theory falls into this category, as it disregards centuries of scientific research and observation. The spread of pseudoscience is often fueled by misinformation, confirmation bias, and the desire to challenge established scientific consensus.

One of the main reasons pseudoscience gains traction is the appeal to conspiracy theories. Flat earth proponents often claim that there is a global conspiracy involving governments, space agencies, and scientists to hide the true shape of the earth. This narrative taps into people's inherent skepticism towards authority and fosters a sense of belonging within a community that questions the status quo.

11.2.2 Misinformation and Confirmation Bias

Misinformation plays a significant role in perpetuating the flat earth theory. With the advent of social media and the ease of sharing information, false claims and distorted facts can quickly spread to a wide audience. Confirmation bias, the tendency to seek out information that confirms preexisting beliefs, further reinforces the spread of misinformation.

Flat earth proponents often cherry-pick evidence that supports their claims while dismissing or ignoring scientific evidence that contradicts their beliefs. They may rely on flawed experiments, misinterpretations of data, or anecdotal evidence to support their arguments. This selective approach to information perpetuates a distorted view of reality and hinders critical thinking.

11.2.3 The Dangers of Ignoring Scientific Consensus

The flat earth movement poses several dangers when it comes to ignoring scientific consensus. Firstly, it undermines the credibility of scientific institutions and the scientific method itself. By rejecting well-established scientific principles, such as the roundness of the earth, the movement fosters a distrust in scientific expertise and promotes a culture of skepticism towards scientific advancements.

Secondly, the spread of pseudoscience and misinformation can have detrimental effects on education. When false information is presented as valid scientific knowledge, it confuses and misleads individuals, particularly students who are in the process of learning about the world. This can hinder their ability to think critically, evaluate evidence, and make informed decisions based on reliable information.

11.2.4 Promoting Critical Thinking and Rationality

To combat the spread of pseudoscience and misinformation, it is crucial to promote critical thinking and rationality. Education plays a vital role in equipping individuals with the tools to evaluate information critically, distinguish between reliable sources and pseudoscience, and understand the scientific method.

Encouraging scientific literacy and teaching individuals how to evaluate evidence, question claims, and think critically can help counter the influence of pseudoscience. By fostering a culture of skepticism that is based on evidence and logical reasoning, we can empower individuals to make informed decisions and contribute to a society that values scientific consensus.

In conclusion, the rise of pseudoscience and misinformation, exemplified by the flat earth theory, poses significant challenges to society. The spread of false information and the rejection of scientific consensus undermine the progress of knowledge and hinder critical thinking. By promoting scientific literacy, encouraging rationality, and addressing the dangers of pseudoscience, we can work towards a society that embraces evidence-based reasoning and values the pursuit of truth.

11.3 Dangers of Ignoring Scientific Consensus

Scientific consensus plays a crucial role in shaping our understanding of the world around us. It represents the collective agreement among experts in a particular field based on extensive research, evidence, and rigorous testing. Ignoring scientific consensus can have significant dangers and consequences, particularly when it comes to beliefs such as the flat earth theory.

11.3.1 Misinformation and Pseudoscience

One of the primary dangers of ignoring scientific consensus is the spread of misinformation and pseudoscience. When individuals reject established scientific principles and instead embrace unfounded beliefs, they open themselves up to a world of misinformation. This can lead to a distorted understanding of reality and a rejection of evidence-based knowledge.

The flat earth theory, for example, is built upon a foundation of pseudoscience and conspiracy theories. By disregarding the overwhelming scientific evidence supporting a round earth, flat earthers perpetuate misinformation that can mislead others and hinder scientific progress. This rejection of scientific consensus can have far-reaching consequences, as it undermines the credibility of scientific research and promotes a culture of skepticism and doubt.

11.3.2 Hindering Progress and Innovation

Ignoring scientific consensus can also hinder progress and innovation in various fields. Scientific consensus is the result of years of research, experimentation, and collaboration among experts. It represents the current state of knowledge and understanding in a particular field. When individuals reject this consensus, they impede the advancement of scientific knowledge and hinder the development of new technologies and solutions.

In the case of the flat earth theory, the rejection of the round earth model limits our ability to explore and understand the world. It hampers advancements in fields such as geodesy, satellite technology, and space exploration. By disregarding scientific consensus, we miss out on the opportunity to expand our knowledge and make groundbreaking discoveries.

11.3.3 Undermining Education and Critical Thinking

Another danger of ignoring scientific consensus is the undermining of education and critical thinking. Scientific consensus is the result of rigorous scientific inquiry and critical evaluation of evidence. It represents the pinnacle of human knowledge and understanding. By dismissing this consensus, individuals undermine the importance of education and critical thinking.

When individuals reject scientific consensus, they often rely on flawed reasoning, logical fallacies, and cherry-picked evidence to support their beliefs. This undermines the principles of critical thinking and rationality, which are essential for evaluating information and making informed decisions. It also sets a dangerous precedent, as it encourages the acceptance of unfounded beliefs without proper scrutiny.

11.3.4 Promoting a Culture of Ignorance

Ignoring scientific consensus can contribute to the promotion of a culture of ignorance. When individuals reject established scientific principles, they create an environment where misinformation and pseudoscience thrive. This can lead to a society that is ill-informed, susceptible to manipulation, and resistant to evidence-based knowledge.

The flat earth theory, for example, has gained traction in recent years through online communities and social media platforms. This has created an echo chamber where individuals reinforce their beliefs and dismiss opposing viewpoints. In such an environment, critical thinking and open-mindedness are often discouraged, further perpetuating ignorance and hindering intellectual growth.

11.3.5 Impeding Global Collaboration

Scientific consensus is not limited to a single country or region; it is a global agreement among experts from various backgrounds and cultures. Ignoring scientific consensus can impede global collaboration and hinder our ability to address pressing global challenges.

In the case of climate change, for instance, the overwhelming scientific consensus supports the idea that human activities are contributing to global warming. By ignoring this consensus, countries and individuals may fail to take necessary actions to mitigate the impacts of climate change. This can have severe consequences for the environment, economy, and future generations.

11.3.6 Promoting Critical Thinking and Rationality

To counter the dangers of ignoring scientific consensus, it is crucial to promote critical thinking and rationality. Encouraging individuals to question and evaluate information critically can help them distinguish between evidence-based knowledge and unfounded beliefs.

Education plays a vital role in promoting scientific literacy and critical thinking skills. By providing individuals with a solid foundation in scientific principles and the scientific method, we can equip them with the tools necessary to evaluate information objectively and make informed decisions.

Furthermore, fostering an environment that values scientific consensus and encourages open dialogue can help combat the spread of misinformation and pseudoscience. By promoting evidence-based knowledge and encouraging respectful discussions, we can create a society that embraces scientific consensus and values the pursuit of truth.

In conclusion, ignoring scientific consensus, such as in the case of the flat earth theory, can have significant dangers and consequences. It promotes the spread of misinformation, hinders progress and innovation, undermines education and critical thinking, promotes a culture of ignorance, and impedes global collaboration. To address these dangers, it is essential to promote critical thinking, scientific literacy, and rationality, while fostering an environment that values scientific consensus and encourages open dialogue.

11.4 Promoting Critical Thinking and Rationality

In a world where misinformation and pseudoscience can easily spread, it is crucial to promote critical thinking and rationality. The flat earth theory is a prime example of how beliefs can be formed without proper evidence or logical reasoning. By encouraging individuals to think critically and evaluate information objectively, we can help combat the spread of false beliefs and promote a more scientifically literate society.

11.4.1 The Importance of Critical Thinking

Critical thinking is the ability to analyze and evaluate information objectively, without bias or preconceived notions. It involves questioning assumptions, examining evidence, and considering alternative explanations. By developing critical thinking skills, individuals can become more discerning consumers of information and less susceptible to misinformation.

Promoting critical thinking is essential because it allows individuals to make informed decisions based on evidence and logical reasoning. It helps people distinguish between reliable sources of information and those that are based on unfounded claims or conspiracy theories. By encouraging critical thinking, we can empower individuals to question and challenge ideas, leading to a more intellectually curious and scientifically literate society.

11.4.2 Teaching Rationality in Education

One of the most effective ways to promote critical thinking and rationality is through education. By incorporating lessons on scientific methodology, logical reasoning, and evidence-based thinking, we can equip students with the tools they need to evaluate information critically.

In science classrooms, teachers can emphasize the importance of empirical evidence and the scientific method. Students can learn how to design experiments, collect data, and draw conclusions based on evidence. By engaging in hands-on activities and experiments, students can develop a deeper understanding of the scientific process and the importance of evidence in forming conclusions.

Additionally, educators can teach students about logical fallacies and cognitive biases. By familiarizing students with common pitfalls in reasoning, they can learn to recognize and avoid them. This knowledge can help students become more discerning consumers of information, enabling them to identify flawed arguments and misleading claims.

11.4.3 Encouraging Skepticism and Curiosity

Promoting critical thinking also involves encouraging skepticism and curiosity. Rather than accepting information at face value, individuals should be encouraged to question and seek evidence to support or refute claims. By fostering a sense of curiosity, individuals are more likely to engage in independent research and seek out reliable sources of information.

Skepticism, when applied appropriately, is a valuable tool in evaluating claims and arguments. It involves questioning the validity of information and demanding evidence to support it. By teaching individuals to be skeptical, we can instill a healthy level of doubt and encourage them to seek out reliable sources of information before forming beliefs.

11.4.4 Debunking Flat Earth Claims

Promoting critical thinking and rationality also involves debunking specific claims made by flat earthers. By addressing these claims with scientific evidence and logical reasoning, we can help individuals understand the flaws in the flat earth theory.

One common claim made by flat earthers is that the horizon always appears flat, suggesting a flat earth. However, this can be easily debunked by understanding the concept of the curvature of the earth. By explaining how the curvature of the earth affects our perception of the horizon, we can demonstrate that the appearance of a flat horizon is consistent with a round earth.

Another claim is that gravity does not exist and that objects simply fall due to density differences. However, by presenting evidence from experiments and observations, we can show that gravity is a fundamental force that explains the motion of objects on earth. By understanding the principles of gravity, individuals can see the flaws in the flat earth argument.

11.4.5 Promoting Scientific Literacy

Ultimately, promoting critical thinking and rationality is closely tied to promoting scientific literacy. Scientific literacy involves understanding the basic principles of science, the scientific method, and the importance of evidence-based thinking. By promoting scientific literacy, we can equip individuals with the knowledge and skills necessary to evaluate claims and make informed decisions.

Scientific literacy can be fostered through various means, including science education, public outreach programs, and media literacy. By providing accessible and accurate scientific information, we can empower individuals to think critically about the world around them. Additionally, promoting media literacy can help individuals navigate the vast amount of information available and identify reliable sources.

11.4.6 The Role of Media and Communication

In today's digital age, media and communication play a significant role in shaping public opinion and beliefs. It is crucial to promote responsible and accurate reporting of scientific information. Journalists and media organizations should strive to present scientific findings in a clear and unbiased manner, avoiding sensationalism or misrepresentation.

Furthermore, scientists and science communicators have a responsibility to engage with the public and communicate their research effectively. By presenting scientific information in an accessible and engaging way, they can help bridge the gap between scientific knowledge and public understanding. This can be done through various mediums, such as documentaries, podcasts, and social media platforms.

11.4.7 The Power of Collaboration and Dialogue

Promoting critical thinking and rationality also involves fostering open dialogue and collaboration. By encouraging respectful discussions and debates, individuals can learn from one another and challenge their own beliefs. This collaborative approach allows for the exchange of ideas and the exploration of different perspectives.

Engaging in constructive conversations with individuals who hold alternative beliefs, such as flat earthers, can be challenging but essential. By approaching these conversations with empathy and understanding, we can create an environment where individuals feel comfortable questioning their beliefs and considering alternative viewpoints. This can lead to a more open-minded and intellectually curious society.

11.4.8 The Responsibility of Individuals

Promoting critical thinking and rationality is a collective responsibility, but it also starts with individuals. Each person has the power to question, seek evidence, and think critically. By taking the initiative to educate ourselves, evaluate information objectively, and engage in respectful discussions, we can contribute to a society that values rationality and evidence-based thinking.

In conclusion, promoting critical thinking and rationality is crucial in combating the spread of false beliefs, such as the flat earth theory. By teaching individuals to think critically, encouraging skepticism and curiosity, debunking false claims, promoting scientific literacy, and fostering open dialogue, we can create a society that values evidence-based thinking and rationality. It is through these efforts that we can embrace the reality of our round earth and navigate the age of information with confidence and clarity.

12 Conclusion

12.1 Summarizing the Evidence

Throughout this book, we have explored the various aspects of the flat earth theory and delved into the scientific evidence that supports the reality of our round earth. In this final section, we will summarize the key evidence that debunks the flat earth theory and solidifies our understanding of the true shape of our planet.

12.1.1 The Shape of Earth

One of the most fundamental pieces of evidence supporting the round earth is the shape of our planet itself. Over centuries, observations and measurements have consistently shown that the earth is spherical. Early theories and observations by ancient civilizations, such as the Greeks and Egyptians, provided initial insights into the roundness of the earth. However, it was not until modern times that we gained a comprehensive understanding of Earth's shape.

Geodetic surveys and measurements conducted around the globe have provided concrete evidence of the earth's curvature. These surveys involve measuring the angles and distances between various points on the earth's surface. The results consistently demonstrate that the earth is curved, with the curvature becoming more pronounced as we move away from the surface.

Furthermore, satellite imagery and space exploration have provided us with a wealth of visual evidence. Images captured by satellites and astronauts clearly show the spherical shape of our planet. These images, coupled with the ability to observe the earth from space, have revolutionized our understanding of Earth's shape and left no room for doubt.

12.1.2 Gravity and Earth's Shape

Gravity plays a crucial role in shaping the earth. The force of gravity pulls matter towards the center of mass, resulting in a spherical shape. This is evident in the way celestial bodies, including planets, moons, and stars, naturally form into spheres due to the influence of gravity.

Experiments and observations on gravity have consistently supported the round earth theory. For instance, the phenomenon of gravity causing objects to fall towards the center of the earth is observed universally. The acceleration due to gravity is also consistent across different locations on the earth's surface, further confirming its spherical shape.

Additionally, the way gravity affects Earth's atmosphere provides further evidence. The atmosphere is held in place by gravity, forming a spherical shape around the planet. The distribution of air pressure and the behavior of weather patterns can be explained by the spherical nature of the earth and the influence of gravity.

12.1.3 Earth's Rotation and the Coriolis Effect

The rotation of the earth on its axis is another piece of evidence that supports its round shape. The Coriolis effect, a phenomenon caused by the rotation of the earth, is observed in various aspects of our daily lives.

The Coriolis effect influences weather patterns, causing the rotation of storms and the formation of distinct wind patterns. This effect is also responsible for the rotation of ocean currents and the direction of projectile motion. These observations align with the round earth theory and provide further evidence of Earth's rotation.

12.1.4 Curvature and Horizon Observations

The curvature of the earth is observable in various ways, further debunking the flat earth theory. Curvature calculations and formulas allow us to predict the amount of curvature we should observe at different distances from the surface.

These calculations align with our observations, confirming the spherical shape of the earth.

In everyday life, we can also observe visible curvature. For example, when standing on a beach, we can see ships gradually disappear over the horizon as they sail away. This phenomenon is only possible if the earth is curved. Similarly, the ability to see farther from higher elevations, such as mountaintops, is a result of the earth's curvature.

Photographic evidence also supports the round earth theory. Images taken from high altitudes or space consistently show the curvature of the earth. These images provide a visual representation of the spherical shape of our planet and leave no room for doubt.

12.1.5 The Overwhelming Scientific Consensus

The evidence presented throughout this book is not isolated or limited to a few select studies. It represents the culmination of centuries of scientific research and observations conducted by countless scientists from around the world. The overwhelming consensus among the scientific community is that the earth is round.

The scientific method, which relies on empirical evidence, peer review, and reproducibility, has been instrumental in solidifying our understanding of the round earth. The flat earth theory, on the other hand, lacks scientific credibility and fails to withstand scientific scrutiny.

12.1.6 Embracing the Reality of Our Round Earth

In conclusion, the evidence overwhelmingly supports the reality of our round earth. From the shape of the planet itself to the influence of gravity, the observations of Earth's rotation, the visible curvature, and the scientific consensus, all point to a spherical earth.

It is crucial for us to embrace this reality and promote scientific literacy. Understanding the shape of our planet is not only a matter of scientific curiosity but also has significant implications for our society, education, and the way we navigate the world. By embracing the reality of our round earth, we can foster critical thinking, rationality, and a deeper appreciation for the wonders of our planet and the universe beyond.

12.2 Addressing Remaining Doubts

While we have explored and debunked many aspects of the Flat Earth theory throughout this book, it is important to address any remaining doubts that may linger in the minds of those who still hold onto this belief. In this section, we will address some of the common doubts and misconceptions that Flat Earthers may have and provide further evidence to support the reality of our round Earth.

12.2.1 The Horizon and Perspective

One of the arguments often put forth by Flat Earthers is the perceived flatness of the horizon. They claim that if the Earth were truly round, the horizon would appear curved. However, this argument fails to take into account the limitations of human perception and the vastness of our planet.

When standing on the ground, the curvature of the Earth is not immediately noticeable due to its large size. The human eye has a limited field of view, and the curvature of the Earth becomes more apparent when observing from higher altitudes or over long distances. Additionally, atmospheric conditions, such as haze or fog, can also affect the visibility of the curvature.

Furthermore, the perception of a flat horizon is also influenced by the way light travels. Light rays from distant objects converge towards the observer, creating the illusion of a flat line. This phenomenon, known as perspective, is the same reason why parallel train tracks appear to converge in the distance.

12.2.2 Gravity and the Flat Earth

Flat Earthers often struggle to explain the force of gravity within their theory. They propose that objects are pulled downwards due to a mysterious force called "universal acceleration." However, this explanation contradicts the well-established scientific understanding of gravity.

Gravity is a fundamental force that attracts objects with mass towards each other. It is responsible for holding celestial bodies in orbit, shaping the structure of the universe, and keeping our feet firmly planted on the ground. The concept of gravity is supported by extensive scientific research, experiments, and observations.

The force of gravity is directly related to the mass of an object. On a flat Earth, gravity would not be able to explain why objects fall towards the ground or why the weight of an object changes depending on its location. The round Earth model, on the other hand, provides a coherent explanation for these phenomena, as gravity pulls objects towards the center of the Earth.

12.2.3 Satellite Imagery and Space Exploration

Flat Earthers often question the authenticity of satellite imagery and space exploration missions. They argue that these images and missions are part of a grand conspiracy to deceive the public. However, this claim is unfounded and lacks substantial evidence.

Satellite imagery has provided us with a wealth of information about our planet, including weather patterns, land formations, and even the effects of human activity on Earth's ecosystems. These images are captured by a network of satellites orbiting the Earth, which transmit data back to Earth for analysis.

Space exploration missions, such as those conducted by NASA and other space agencies, have provided us with invaluable knowledge about our solar system and the universe. These missions have allowed us to study other planets, moons, and even distant galaxies. The technology and data gathered from these missions have revolutionized our understanding of the cosmos.

The notion that all space agencies and thousands of scientists worldwide are part of a vast conspiracy is highly implausible. It would require an unprecedented level of coordination and secrecy, which is simply not feasible. Furthermore, the evidence provided by satellite imagery and space exploration missions aligns with the round Earth model and supports the scientific consensus.

12.2.4 Debunking the Ice Wall Myth

One of the most persistent myths associated with the Flat Earth theory is the existence of an "Ice Wall" surrounding the Earth. Flat Earthers claim that this wall prevents us from reaching the edge of the Earth and discovering the truth. However, this notion is purely speculative and lacks any substantial evidence.

Explorations of the Antarctic region have revealed that there is no physical barrier or wall encircling the Earth. The Antarctic continent is a vast landmass covered in ice, but it does not form a continuous wall around the planet. Expeditions and scientific research conducted in Antarctica have provided ample evidence to support the round Earth model.

Furthermore, the circumnavigation of the Earth by ships and airplanes is a testament to the absence of an impenetrable ice wall. If such a wall existed, it would be impossible to travel from one side of the Earth to the other without encountering it. The ability to circumnavigate the globe is a clear indication that the Earth is round.

In conclusion, the remaining doubts and misconceptions surrounding the Flat Earth theory can be addressed and debunked through scientific evidence, observations, and logical reasoning. The round Earth model, supported by centuries of research and exploration, provides a comprehensive and coherent understanding of our planet's shape and the forces that govern it. Embracing the reality of our round Earth is not only essential for scientific literacy but also for fostering critical thinking and rationality in an age of misinformation.

12.3 The Importance of Scientific Literacy

Scientific literacy is a crucial aspect of our modern society. It empowers individuals to understand and evaluate scientific information, enabling them to make informed decisions and navigate the complexities of the world around them. In the context of the flat earth theory, scientific literacy plays a vital role in debunking misconceptions and promoting rational thinking.

12.3.1 Understanding the Scientific Method

Scientific literacy begins with an understanding of the scientific method, the systematic approach used by scientists to investigate and understand the natural world. This method involves making observations, formulating hypotheses, conducting experiments, analyzing data, and drawing conclusions. By grasping the scientific method, individuals can discern between reliable scientific evidence and baseless claims.

12.3.2 Evaluating Sources and Discerning Misinformation

In an era of abundant information, scientific literacy equips individuals with the skills to critically evaluate sources and discern misinformation. It enables them to differentiate between peer-reviewed scientific research and pseudoscience, conspiracy theories, or personal anecdotes. By understanding the importance of rigorous scientific scrutiny, individuals can avoid falling prey to misleading information.

12.3.3 Promoting Skepticism and Critical Thinking

Scientific literacy fosters skepticism and critical thinking, encouraging individuals to question and challenge ideas. It teaches them to examine evidence, consider alternative explanations, and weigh the credibility of different arguments. By cultivating a healthy skepticism, individuals can avoid accepting claims at face value and instead seek evidence-based explanations.

12.3.4 Debunking Flat Earth Misconceptions

Scientific literacy is particularly crucial in debunking misconceptions surrounding the flat earth theory. By understanding the scientific evidence that supports a round earth, individuals can counter the flawed arguments put forth by flat earthers. They can recognize the fallacies in claims such as the supposed flatness of the horizon or the existence of an ice wall encircling the earth.

12.3.5 Appreciating the Beauty of Scientific Discoveries

Scientific literacy allows individuals to appreciate the beauty and wonder of scientific discoveries. It enables them to understand the vastness of our universe, the intricacies of our planet, and the remarkable achievements of human exploration. By embracing scientific literacy, individuals can develop a sense of awe and curiosity about the world, fostering a lifelong love for learning and discovery.

12.3.6 Nurturing a Respect for Expertise

Scientific literacy also nurtures a respect for expertise and the value of scientific consensus. It recognizes that scientific knowledge is built upon the collective efforts of experts in their respective fields. By acknowledging the expertise of scientists, individuals can appreciate the rigorous process of peer review, the importance of reproducibility, and the significance of scientific consensus in shaping our understanding of the world.

12.3.7 Addressing Misconceptions and Skepticism

Scientific literacy equips individuals with the tools to address misconceptions and skepticism effectively. By engaging in respectful and evidence-based discussions, individuals can help dispel myths and promote accurate scientific information. They can provide clear explanations, share reliable sources, and encourage others to critically evaluate their beliefs.

12.3.8 Fostering a Culture of Scientific Inquiry

Scientific literacy is not only important for individuals but also for society as a whole. It fosters a culture of scientific inquiry, where evidence-based thinking and rationality are valued. By promoting scientific literacy, we can encourage the next generation to pursue careers in science, contribute to scientific advancements, and address the challenges facing our world.

12.3.9 Empowering Individuals to Make Informed Decisions

Ultimately, scientific literacy empowers individuals to make informed decisions that affect their lives and the world around them. It enables them to understand complex issues such as climate change, vaccination, and technological advancements. By being scientifically literate, individuals can actively participate in discussions, advocate for evidence-based policies, and contribute to the betterment of society.

12.3.10 Embracing the Reality of Our Round Earth

In conclusion, scientific literacy is of utmost importance in debunking the flat earth theory and promoting rational thinking. It equips individuals with the skills to evaluate scientific information, discern misinformation, and critically analyze claims. By embracing scientific literacy, we can appreciate the wonders of our round earth, contribute to scientific progress, and make informed decisions that shape our future.

12.4 Moving Forward in the Age of Information

In the age of information, where knowledge is readily accessible at our fingertips, it is crucial that we continue to promote scientific literacy and critical thinking. The prevalence of misinformation and conspiracy theories, such as the flat earth theory, highlights the importance of educating ourselves and others about the reality of our round earth. Moving forward, we must address the remaining doubts and misconceptions surrounding this topic, while also promoting rationality and scientific understanding.

12.4.1 Addressing Misconceptions and Doubts

Despite the overwhelming scientific evidence supporting the round earth model, there are still individuals who hold onto the belief in a flat earth. It is essential to address their misconceptions and doubts in a respectful and informative manner. By engaging in open and honest discussions, we can provide them with the necessary information to challenge their beliefs and encourage critical thinking.

One common misconception among flat earthers is the idea that the curvature of the earth is not visible in everyday life. They argue that if the earth were truly round, we would be able to see the curvature from any vantage point. However, this misconception can be easily debunked by understanding the limitations of human perception. The curvature of the earth is indeed visible, but it becomes more apparent at higher altitudes or when observing large bodies of water.

Another doubt often raised by flat earthers is the belief in a vast ice wall surrounding the earth. They argue that this ice wall acts as a barrier, preventing us from reaching the edge of the flat earth. However, scientific exploration and research in the Antarctic region have consistently disproven this claim. The Antarctic continent is not an ice wall encircling the earth but a vast landmass covered in ice. Expeditions and satellite imagery have revealed the true nature of Antarctica, showcasing its diverse landscapes and wildlife.

12.4.2 Promoting Critical Thinking and Rationality

To combat the spread of misinformation and conspiracy theories, it is crucial to promote critical thinking and rationality. By encouraging individuals to question and evaluate the evidence presented to them, we can foster a society that values scientific literacy and skepticism. Critical thinking allows us to analyze information objectively, separate fact from fiction, and make informed decisions based on evidence.

Educational institutions play a vital role in promoting critical thinking skills. By incorporating scientific literacy into the curriculum, students can develop the necessary tools to evaluate information critically. Teaching students how to analyze data, question assumptions, and understand the scientific method empowers them to navigate the vast amount of information available in the digital age.

Furthermore, promoting rationality involves addressing cognitive biases and belief systems that may hinder the acceptance of scientific evidence. Confirmation bias, for example, is a common cognitive bias that leads individuals to seek out information that confirms their preexisting beliefs while ignoring contradictory evidence. By raising awareness about these biases and encouraging individuals to challenge their own beliefs, we can create a more rational and open-minded society.

12.4.3 Embracing Scientific Literacy

Scientific literacy is the foundation upon which we can build a society that values evidence-based knowledge. It involves understanding the scientific method, evaluating scientific claims, and recognizing the importance of peer-reviewed research. By promoting scientific literacy, we equip individuals with the tools to critically evaluate information and make informed decisions.

In the age of information, it is crucial to teach individuals how to discern reliable sources of information from unreliable ones. The internet has provided us with a wealth of knowledge, but it has also given rise to misinformation and pseudoscience. By teaching individuals how to evaluate the credibility of sources, we can empower them to make informed decisions based on reliable information.

Scientific literacy also involves understanding the limitations of our own knowledge and being open to new discoveries and advancements. Science is a dynamic field that is constantly evolving, and it is important to embrace new evidence and revise our understanding accordingly. By fostering a culture of curiosity and lifelong learning, we can ensure that scientific literacy remains a cornerstone of our society.

12.4.4 The Role of Technology and Communication

Technology and communication play a crucial role in disseminating scientific knowledge and combating misinformation. The internet and social media platforms have the potential to reach a wide audience and provide access to reliable scientific information. Scientists, educators, and science communicators can utilize these platforms to share accurate information, debunk myths, and engage with the public.

However, it is important to recognize the challenges that come with the digital age. The rapid spread of misinformation and the echo chamber effect, where individuals are exposed only to information that aligns with their beliefs, pose significant obstacles. To overcome these challenges, it is essential to develop effective science communication strategies that are engaging, accessible, and tailored to different audiences.

Collaboration between scientists, educators, and communicators is key to ensuring that accurate scientific information reaches the public. By working together, we can bridge the gap between scientific research and public understanding, fostering a society that values evidence-based knowledge and critical thinking.

Conclusion

In conclusion, moving forward in the age of information requires us to address the remaining doubts and misconceptions surrounding the round earth model. By promoting critical thinking, rationality, and scientific literacy, we can combat the spread of misinformation and conspiracy theories. Embracing technology and effective science communication strategies will allow us to disseminate accurate scientific information and engage with the public. As we continue to uncover the truth about our round earth, it is essential that we embrace scientific understanding and promote a society that values evidence-based knowledge.